FUZZY SET APPROACH TO MULTIDIMENSIONAL POVERTY MEASUREMENT

ECONOMIC STUDIES IN INEQUALITY, SOCIAL EXCLUSION AND WELL-BEING

This series will publish volumes that go beyond the traditional concepts of consumption, income or wealth and will offer a broad, inclusive view of inequality and well-being. Specific areas of interest will include Capabilities and Inequalities, Discrimination and Segregation in the Labor Market, Equality of Opportunities, Globalization and Inequality, Human Development and the Quality of Life, Income and Social Mobility, Inequality and Development, Inequality and Happiness, Inequality and Malnutrition, Income and Social Mobility, Inequality in Consumption and Time Use, Inequalities in Health and Education, Multidimensional Inequality and Poverty Measurement, Polarization among Children and Elderly People, Social Policy and the Welfare State, and Wealth Distribution.

Volume 1
de Janvry, Alain and Kanbur, Ravi
Poverty, Inequality and Development: Essays in Honor of Erik Thorbecke

Volume 2
Duclos, Jean-Yves and Araar, Abdelkrim
Poverty and Equity: Measurement, Policy and Estimation with DAD

Volume 3
Lemmi, Achille and Betti, Gianni
Fuzzy Set Approach to Multidimensional Poverty Measurement

FUZZY SET APPROACH TO MULTIDIMENSIONAL POVERTY MEASUREMENT

edited by

Achille Lemmi and Gianni Betti

Library of Congress Control Number: 2006925109

ISBN:10: 0-387-34249-4 (Printed on acid-free paper)
ISBN-13: 978-0387-34249-8

e-ISBN-10: 0-387-34251-6
e-ISBN-13: 978-0387-34251-1

Printed in the United States of America.

9 8 7 6 5 4 3 2 1

springer.com

Table of Contents

List of contributing authors

Berenger, Valerie, Assistant Professor of Economics

Faculty of law and Economics, CEMAFI, University of Nice-Sophia Antipolis

Av. Doyen Trotabas, 06050 Nice Cedex 01, France

Valerie Berenger is Assistant Professor of Economics at the University of Nice-Sophia-Antipolis and is a member of the CEMAFI (Centre d'Etudes en Macroéconomie et Finance Internationale). She was awarded a PhD in economics in 1998 on pension financing schemes. Her research interests are in social economics, in general, with special emphasis on pension financing schemes and social welfare policies. In addition to these studies, her current research examines multidimensional poverty measurement methods.

Betti, Gianni, Associate Professor of Economic Statistics

Dipartimento Metodi Quantitativi, University of Siena

P.zza S. Francesco, 8, 53100 Siena, Italy

Gianni Betti is Associate Professor in Economic Statistics and member of the C.R.I.DI.RE. (Centro Interdipartimentale di Ricerca sulla Distribuzione del Reddito), University of Siena, Italy. He has worked on several projects for the World Bank and European Commission, and has been closely involved with the development of the EU Statistics on Income and Living Conditions (EU-SILC).

Celestini, Franck, Assistant Professor

Faculty of Science, University of Nice-Sophia Antipolis, LPMC

CNRS UMR 6622, Parc Valrose, 06108 Nice Cedex 02, France

Franck Celestini was awarded a PhD in condensed matter Physics in 1995 at the University of Nice-Sophia Antipolis. He was then a post-doctoral associate researcher at the International School of Advanced Studies in Trieste (Italy). He obtained a permanent position at the University of Marseille in 1997 and is currently Assistant Professor at the University of Nice (LPMC-CNRS). His main research activity is on the modelisation and simulation of condensed matter and statistical physic systems.

Chakravarty, R. Satya, Professor of Economics

Economic Research Unit, Indian Statistical Institute

203, B. T. Road, Kolkata, 700 108, India

Satya R. Chakravarty is a Professor of Economics at the Indian Statistical Institute, Kolkata, India. He has published articles in Econometrica, Journal of Economic Theory, Games and Economic Behavior, International Economic Review, Economic Theory, Social Choice and Welfare, Journal of Development Economics, Canadian Journal of Economics, Mathematical Social Sciences , Economics Letters and Theory and Decision; as well as book published by Springer-Verlag. His major research interests are: measurement of inequality, well-being, poverty, deprivation etc.; and cooperative games.

Cheli, Bruno, Associate Professor of Economic Statistics
Dipartimento di Statistica e Matematica applicata all'Economia
University of Pisa, Via Ridolfi 10, 56124 Pisa, Italy
Bruno Cheli is Associate Professor in Economic Statistics at the University of Pisa and member of the C.R.I.DI.RE., University of Siena. His main research topics are the measurement and analysis of poverty, living conditions and sustainable development.

Chiappero Martinetti, Enrica, Associate Professor in Economics
Dipartimento Economia Pubblica e Territoriale, University of Pavia
Strada Nuova 65, 27100 Pavia, Italy
Enrica Chiappero Martinetti is Associate Professor in Economics at the Faculty of Political Science, University of Pavia (Italy). Her main research interests are in the field of poverty and inequality theory and measurement, fuzzy set theory applied to poverty and well-being analysis, human development, capability approach and multidimensional analysis of well-being. She is Vice-President of the Human Development Capability Association.

D'Agostino, Antonella, Researcher in Economic Statistics
Dipartimento di Statistica e Matematica per la Ricerca Economica, University of Naples "Parthenope"
Via Medina, 40, 80133 Napoli, Italy
Antonella D'Agostino is Researcher in Economic Statistics. She has collaborated with Eurostat in some research projects on poverty and living conditions and currently her research interest fields are concentrated above all in the youth labour market.

Deutsch, Joseph, Associate Professor in Economics
Department of Economics, Bar-Ilan University
52900 Ramat-Gan, Israel
Joseph Deutsch has been an Associate Professor in the Department of Economics at Bar-Ilan University since 1984 and has published over 50 scientific papers in international academic journals. He served as the Chair of The Interdisciplinary Department of Social Sciences and for several years was a visiting professor in the Economic Department of the University of Pennsylvania. Currently he also serves in the Board of Editors of the Journal of Economic Inequality, the Atlantic Economic Journal and the International Journal of Business.

Duma, Viorica
National Institute of Statistics
16, Libertatii Ave., Sector 5, Bucharest, Romania
Viorica Duma, National Institute of Statistics, Romania. She is a mathematician with working experience in the areas of sampling surveys, poverty analyses, programming in SAS language. She is co-author for reports, evaluations and articles on poverty evaluation, living conditions, child labour, sampling design, models building. She is currently head of sampling Office for Social Statistics.

Fustier, Bernard, Professor of Mathematics,
Università di Corsica, Pasquale Paoli, Corse, France

Bernard Fustier, former member of the spatial economists team from the University of Burgundy, has been teaching economics and econometrics at the University of Corsica (and for some time spatial analysis at Aix-en-Provence and Nice Universities). Being well aware that all is not measurable, he has become interested in softer forms of mathematics and has written several articles dealing with multicriteria aid for decision and fuzzy logic.

Lemmi, Achille, Full Professor of Economic Statistics
Dipartimento Metodi Quantitativi, University of Siena
P.zza S. Francesco, 8, 53100 Siena, Italy

Achille Lemmi is Full Professor in Statistics and Economics and Director of C.R.I.DI.RE. at the University of Siena, Italy. He is Associate Editor of Statistics in Transition and member of the Editorial Board of Journal of Economic Inequality. His life-long research interest has been the study of income distribution, inequality and poverty.

Menirav, Ehud
The Eitan Berglas School of Economics, Tel-Aviv University, Israel

Ehud Menirav received his Ph.D in Economics from The Eitan Berglas School of Economics, Tel-Aviv University in 2005. He has an MA in Business Economics and a BA in Economics and Business Administration, both from Bar-Ilan University, Israel. His expertise is in Poverty Measurement, Political Economy and Public Choice.

Miceli, David
Département des Finances, Organisation et systèmes d'information
29, rue du Stand, Case postale 3937, CH-1211 Genève 3, Switzerland

David Miceli studied poverty measurement in his PhD thesis at the University of Geneva. He now works for the Department of Finance of the Canton Geneva in Switzerland, where he is in charge of revenue forecasting and tax simulations.

Molnar, Maria, Senior Researcher and Professor of Statistics
Institute of National Economy – Romanian Academy
Casa Academiei - Bucuresti sector 5, Calea 13 Septembrie nr.13, Romania

Maria Molnar has a PhD in Economics. She is a senior researcher at the Institute of National Economy – Romanian Academy and professor of statistics at the Ecological University of Bucharest. Her main areas of specialization are social economics and social statistics, income distribution, poverty and social protection.

Neri, Laura, University Researcher in Statistics
Dipartimento Metodi Quantitativi, University of Siena
P.zza S. Francesco, 8, 53100 Siena, Italy

Laura Neri is University Researcher in Statistics and member of the C.R.I.DI.RE. at the University of Siena, Italy. She has been involved in several projects, regarding income distribution and living conditions, with the World Bank and the European Commission.

Qizilbash, Mozaffar, Lecturer in Economics
School of Economics, University of East Anglia
Norwich, NR4 7TJ, United Kingdom
Mozaffar Qizilbash teaches economics at the University of East Anglia. His research covers a broad range of topics which fall at the fuzzy borderline between economics and philosophy. These include well-being, poverty and vagueness. He is currently editor of the Journal of Human Development and editor-at-large of the Human Development and Capability Association.

Panduru, Filofteia, Co-ordinator of Social Statistics
National Institute of Statistics
16, Libertatii Ave., Sector 5, Bucharest, Romania
Filofteia Panduru has a PhD in Economics. She is the co-ordinator of social statistics in Romania as general director in the National Institute for Statistics. Her main areas of specialization are labour force, living conditions, poverty and social inclusion.

Panek, Tomasz, Full Professor in Economics
Institute of the Statistics and Demography, Warsaw School of Economics
Al. Niepodległości 162, 02-554 Warsaw, Poland
Tomasz Panek, Full Professor and vice director of the Institute of the Statistics and Demography of the Warsaw School of Economics (Poland). His research interests include mainly the issues of living conditions of households, in particular the field of poverty and social inequalities. He is the co-author of the Social Diagnosis panel research project focused on living conditions and quality of life of households in Poland. He is deputy chairman of the Council for Social Monitoring and the co editor of Statistics in Transition.

Silber, Jacques, Professor of Economics
Dept. of Economics, Bar-Ilan University
52900 Ramat-Gan, Israel
Jacques Silber is Professor of Economics at Bar-Ilan University, Israel. He holds a Ph.D. (1975) in Economics from the University of Chicago. He is the editor of a Handbook on Income Inequality Measurement (Kluwer), the author with Y. Flückiger of The Measurement of Segregation and Discrimination in the Labor Force (Physica-Verlag) and the Editor-in-Chief of the Journal of Economic Inequality (Springer). He has published more than 80 articles in international academic journals.

Vasile, Andreea
National Institute of Statistics

16, Libertatii Ave., Sector 5, Bucharest, Romania

Andreea Vasile, National Institute of Statistics (NIS), Romania: co-ordinator of the living conditions statistics department in the institute; is involved in designing and implementing surveys on population living conditions; co-author of studies and reports regarding social inclusion and poverty; Ph.D. student in economics.

Verma, Vijay, Professor in Statistics
Dipartimento Metodi Quantitativi, University of Siena
P.zza S. Francesco, 8, 53100 Siena, Italy

Vijay Verma is Professor at the University of Siena, Italy, and Director of the International Social Research (ISR) Ltd. in London. He has worked for the development of statistics internationally, including with the United Nations, the International Labour Office and Eurostat, and has been closely involved with two major survey research programmes in the EU: the European Community Household Panel (ECHP) and EU-SILC.

Vero, Josiane, Researcher
Céreq, Département 'Production et Usages de la Formation Continue'
10 place de la Joliette, BP 21 321 Marseille cedex 02, France

Josiane Vero is Researcher at the Céreq, a French public body which is involved in the area of training and employment. She is an economist. Her main research interests are the application of theories of justice to welfare, education, training and the labour market. She has over ten years of experience in the measurement of poverty and inequality.

Introduction

Achille Lemmi, Gianni Betti

Dipartimento Metodi Quantitativi, University of Siena

One of the results that nowadays seems to be well-established and shared by scholars in international literature is the definition of poverty and deprivation or, more generally, of the hardships facing individuals and families in multidimensional terms. In fact it is not only theoretical socio-economic research which focuses on this aspect, following up and refining the arguments and formulations put forward by precursors of this approach (Townsend 1979; Sen 1985; Desai and Shah 1988; Bourguignon and Chakravarty 2003), but also several important public statistical agencies which welcome the above-mentioned theoretical arguments and transform them into empirical measures and analyses. The most relevant application of such a method has been conducted by Eurostat (2002) in the second Social Report in Poverty and Social Exclusion in the European Union.

Very recently (August 2005) at an International Conference organized in Brasilia[1] by the UNDP International Poverty Centre on "Many Dimensions of Poverty", the multidimensional nature of poverty was examined from a socio-economic point of view, and also analyzed in an anthropological, psychological, juridical and institutional framework.

Space and attention have also been dedicated to multidimensional poverty measurement in both developing and developed countries, where it has not only been studied theoretically, but also applied empirically. Such quantitative aspects which seem to be strongly representative of the most recent and widespread contributions in international literature all have the same problem of how to deal with the basic information which is very vast and dispersed and therefore have the need for aggregation in synthetic measures.

The two most important types of approach are as follows: i) one which is first of all represented by theoretical construction based on suitable and coherent logical reference models; ii) a second which builds on multivariate statistical methodology (discriminant analysis, factor analysis, cluster analysis, correspondence analysis) and attempts to aggregate, within rea-

[1] For draft versions of presented papers go to web site http://www.undp-povertycentre.org/md-poverty/index.htm

sonable dimensions, the basic information dispersed in considerably extensive indicator vectors.

An approach that belongs to category i) can be derived from the mathematical theory of the fuzzy sets, proposed by Zadeh (1965) and developed by Dubois and Prade (1980). Following this approach the so called Totally Fuzzy and Relative (TFR) method has been proposed by Cheli and Lemmi (1995) starting from an idea of Cerioli and Zani (1990). Further contributions (either for implementing the robustness and for considering the dynamic of poverty) have strongly improved the initial proposal and many applications have been undertaken in several economic and political realities at international, national and regional level, for developed, developing and less developed countries.

An important characteristic of the multidimensional poverty measures derived from the fuzzy approach consists of its particular suitability for small area poverty estimation; in such a sense the European Commission (2005) has adopted the fuzzy approach in the report on "Regional Indicators to reflect social exclusion and poverty".

Moreover the Fuzzy poverty approach shows an intrinsic strong capacity when dynamic statistical information is involved; it is a matter of fact that longitudinal fuzzy poverty estimates allow us to analyze the duration and the intensity of poverty together. Finally, several Authors (Clark and Qizilbash 2002; Chiappero-Martinetti 2000; Lelli 2001) have shown the strict consistency of the fuzzy approach with the theory of the Nobel Prize Amartya Sen on the Capability approach.

It is therefore evident that the international literature on poverty analysis shows an increasing interest on the method mentioned above, but the contributions are wide spread and often not well linked to each other.

The aim of the Book is to provide the interested reader with an organic, consistent and fully comprehensive overview of the fuzzy approach. For reaching such an aim, the Book is divided into two Sections; the first devoted to the theoretical, philosophical, mathematical and statistical aspects, and the second containing further developments based on empirical analyses conducted on actual cross sectional and longitudinal data sets.

Since the fuzzy approach is very innovative with respect to the traditional socio-economic literature, an introduction to its philosophical fundaments and to its mathematical background seems to be appropriate for allowing the reader to fully comprehend the basic content of the Book.

In Chapter 1 Mozaffar Qizilbash discusses the "Philosophical Accounts of vagueness, Fuzzy Poverty Measures and Multidimensionality". The justification for using fuzzy set theory to study poverty is derived from the vague predicate nature of the phenomenon itself.

The Author considers vagueness within the framework of the most diffused philosophical accounts (epistemic view, degree theory and supervaluationism) for defining in correct and detailed terms the foundation of vagueness itself.

At first sight the degree theory appears to be the most appropriate philosophical background since fuzzy set theory is one particular form of such theory. Anyway, an alternative framework inspired by supervaluationism also provides intuitive interpretation of fuzzy poverty estimates, also in the multidimensional case. Further extensions to the longitudinal dimension of poverty allow for the vagueness of predicates such as "extreme" and "chronic" poverty.

The second Chapter of the Book deals with "The Mathematical Framework of Fuzzy Logic" and its application to economics. The Author, Bernard Fustier exposes the principal of graduality, dealing with the counter position of the fuzzy set logic to the classical logic, characterized by the strict opposition true / false, according to the notions of fuzzy proposition, fuzzy set and fuzzy number. The connectors (links) in the fuzzy logic are defined with particular reference to the triangular norms and co-norms and to the Zadeh's operators (max, min and complement to one). The fuzzy optimization according to Bellmann and Zadeh follows and the fuzzy evaluation completes the mathematics of fuzzy set.

Satya R. Chakravarty is the Author of Chapter 3 of the Book on "An Axiomatic Approach to Multidimensional Poverty Measurement via Fuzzy Sets". Since the theory of fuzzy sets enables one to talk of imprecisions in a meaningful way, a proposal to extend multidimensional poverty measurement to a fuzzy environment is attempted. Suitable fuzzy analogues are proposed for various standard index properties. Possible fuzzy indices associated with some multidimensional poverty indices are also proposed.

In Chapter 4 Ehud Menirav examines the "Convergence of Various Unidimensional Approaches", referring to the experience of Israel in 1997. The Author performs a sensitivity analysis based on a comparison of 48 distributions derived from the *Household Expenditures Survey*; those distributions vary according to most relevant analytic elements of the unidimensional monetary approach (well-being variable, equivalence scale, weighting system). The most relevant conclusion is that poverty measures are very sensitive to the choices described above. Therefore the unidimensional monetary approach appears to be inadequate in correctly interpreting the complexity of the poverty phenomenon.

Enrica Chiappero Martinetti in Chapter 5 on "Capability approach and fuzzy set theory: description, aggregation and inference issues", underlines the already mentioned attitude of the fuzzy approach for representing the graduality rather than the dualisms in defining poverty. In particular, this is in line with Amartya Sen's capability approach. It goes beyond a merely multidimensional reference framework with a gradual and non dichotomous interpretation of the complex phenomenon of poverty.

The most updated and complete version of the so called Totally Fuzzy and Relative (TFR) approach to the measurement of poverty is presented in Chapter 6 by Gianni Betti, Bruno Cheli, Achille Lemmi and Vijay Verma on "Multidimensional and longitudinal poverty: an integrated fuzzy approach". Two are the important aspects developed by the Authors: (i) the choice of membership functions i.e. quantitative specification of individuals' or households' degrees of poverty and deprivation, given the level and distribution of income and other aspects of living conditions of the population; and (ii) the choice of rules for the manipulation of the resulting fuzzy sets, rules defining their complements, intersection, union and aggregation. In relation to (i), the Authors propose an "Integrated Fuzzy and Relative" approach, showing a relationship with the Lorenz curve and the Gini coefficient.

The second Section of the Book mainly deals with applications of multidimensional fuzzy poverty analysis. The Countries examined are France, Israel, Romania, Switzerland, Poland and the United Kingdom, which constitute a sufficiently different and wide range of case studies; the examined time periods are quite recent and the underlying methodologies are either derived from the TFR approach or based on some other original contribution derived from the Fuzzy Sets theory.

Four Chapters of the second part of the Book refer to empirical studies based on cross sectional data sets. Three Chapters are devoted to the longitudinal aspects of multidimensional fuzzy poverty analysis. In each of these Chapters original contributions are present.

Valérie Berenger and Franck Celestini in Chapter 7 on "French Poverty Measures using Fuzzy Set Approaches", perform a sensitivity analysis according to the number of different empirical variables for the robustness of the fuzzy poverty index. On the basis of such analysis they propose a new method (based on the TFR approach) in order to deal with the possibility of extracting a law from multidimensional poverty scores analogous to the power law identified by Pareto from income data.

Joseph Deutsch and Jacques Silber in Chapter 8 present "The "Fuzzy Set" Approach to Multidimensional Poverty Analysis: Using the Shapley Decomposition to Analyze the Determinants of Poverty in Israel". The Chapter compares three "fuzzy set approaches" to multidimensional poverty measurement, the Totally Fuzzy Absolute (TFA), the TFR and the Vero and Werquin approaches on the basis of the 1995 Israeli Census. First various cross-tables are given to show the impact of different factors on poverty. Then logit regressions are used to determine, *ceteris paribus*, the exact impact of these factors. Finally results based on a Shapley decomposition are presented to find out which of these determinants are really important.

Chapter 9 on "Multidimensional fuzzy set approach poverty estimates in Romania" by Maria Molnar, Filofteia Panduru, Andreea Vasile and Viorica Duma, consists of an empirical poverty estimation comparing two official unidimensional approaches with to a multidimensional fuzzy approach.

The Authors show how multidimensional analysis of poverty allows a more shading estimation of poverty degree, and outlines the core poverty profile, characterized by more symptoms and dimensions.

Analogous results are obtained by David Miceli in Chapter 10 on "Multidimensional and Fuzzy Poverty in Switzerland". In fact, using a multidimensional fuzzy approach, not only helps in giving a more complete picture of living conditions, but also provides a more accurate picture of poverty which is as near as possible to what may be perceived by simply observing reality.

In Chapter 11 Josiane Vero presents "A comparison of poverty according to primary goods, capabilities and outcomes. Evidence from French school leaver's surveys". In defining the multidimensional framework of poverty the Author considers three different ethical styles with particular reference to: i) social primary goods according to the Rawles' theory of justice; ii) social outcomes following Fleurbaey's definition; and iii) basic capabilities according to the theory of Sen. The three approaches have been tested on the CEREQ French data set; the most important conclusion is that different definitions of poverty identify different poverty profiles or segments of the population.

Tomasz Panek has presented "Multidimensional Fuzzy Relative Poverty Dynamic Measures in Poland" in Chapter 12. The estimation of poverty dynamics via the TFR method, is experimented on a Polish panel survey conducted from 1996 to 1999, during the transition of the Country to the market economy. Such a contribution gives an account of the weakness of the traditional approach, compared with the fuzzy approach, when several movements from poverty and non poverty (and vice-versa) occur.

Finally in Chapter 13 Gianni Betti, Antonella D'Agostino and Laura Neri present "Modelling fuzzy and multidimensional poverty measures in the United Kingdom with variance components panel regression". This last contribution of the Book deals with a panel regression approach for measuring the degree of poverty and living condition; this model is estimated considering as response variables, two fuzzy poverty measures, one based on the monetary indicator, and the other one based on multidimensional indicators. The empirical analysis is conducted on the basis of the waves 1-

12 of the British Household Panel Study, one of the most authoritative and complete longitudinal survey in Europe.

Referring to a non conventional and innovative use of regression models with variance components and TFR poverty estimates, the Authors show how those fuzzy measures can overcome a typical limitation of the traditional unidimensional approach: overestimation of individual mobility near the poverty line.

The topics presented here probably do not completely exhaust the complexity of the arguments contained in the fuzzy approach to the measurement of poverty. The Book should however allow the reader to become more familiar with additional, if not alternative, methods for analyzing poverty as well as living conditions in a correct and systematic way.

Acknowledgments

The Editors would like to thank Jacques Silber for having drawn their attention to the need for and the importance of organically collecting the most relevant topics on the fuzzy set approach to the measurement of poverty.

The Editors are particularly grateful to the Authors of the various Chapters for their commitment and for their desire to contribute to the creation of the Book.

Particular thanks go to colleagues Laura Neri and Vijay Verma for their constant support and precious suggestions.

Many thanks also to Francesca Ballini and Janet Donovan for looking after the editorial and linguistic aspects of the Book.

References

Bourguignon F, Chakravarty SR (2003) The Measurement of Multidimensional Poverty. Journal of Economic Inequality 1:25-49

Cerioli A, Zani S (1990) A Fuzzy Approach to the Measurement of Poverty. In: Dagum C, Zenga M (eds) Income and Wealth Distribution, Inequality and Poverty, Studies in Contemporary Economics, Springer Verlag, Berlin, pp 272-284

Cheli B, Lemmi A (1995) A "Totally" Fuzzy and Relative Approach to the Multidimensional Analysis of Poverty. Economic Notes 24:115-134

Chiappero-Martinetti E (2000) A multidimensional assessment of well-being based on Sen's functioning approach. Rivista Internazionale di Scienze Sociali 108:207-239

Clark D, Qizilbash M (2002) Core poverty and extreme vulnerability in South Africa. The Economics Research Centre, School of Economic and Social Studies, University of East Anglia, Discussion Paper 2002/3
Desai M, Shah A (1988) An Econometric Approach to the Measurement of Poverty. Oxford Economic Papers 40:505-522
Dubois D, Prade H (1980) Fuzzy Sets and Systems: Theory and Applications. Academic Press, New York
European Commission (2005) Regional Indicators to reflect social exclusion and poverty. Employment and Social Affairs DG, Brussels
Eurostat (2002) European Social Statistics: Income, Poverty and Social Exclusion: 2nd Report. Luxembourg: Office for Official Publications of the European Communities
Lelli S (2001) Factor Analysis vs. Fuzzy Sets Theory: Assessing the Influence of Different Techniques on Sen's Functioning Approach. Discussion Paper Series DPS 01.21, November 2001, Center for Economic Studies, Catholic University of Louvain
Sen AK (1985) Commodities and capabilities. North Holland, Amsterdam
Townsend P (1979) Poverty in the United Kingdom. Allen Lane, Harmondsworth
Zadeh LA (1965) Fuzzy sets. Information and Control 8:338-353

1 Philosophical Accounts of Vagueness, Fuzzy Poverty Measures and Multidimensionality

Mozaffar Qizilbash[1]

School of Economics, University of East Anglia

1.1 Introduction

There are a number of phenomena studied by economists and other social scientists which involve vague predicates. Yet economists have not developed any explicit methodology to deal with vagueness. Vagueness is, nonetheless, addressed in the growing literature on fuzzy set theoretic poverty measures (or "fuzzy poverty measures", for short). The emergence of this literature is important not only because it has provided insight into the nature of poverty, but also because of its pioneering contribution at the methodological level. In spite of the increasing use of fuzzy poverty measures there has been limited foundational discussion of vagueness and poverty measurement. In this Chapter, I extend earlier work (Qizilbash 2003) and relate the philosophical literature on vagueness to the literature on poverty measurement. I explain why "poor" is regarded as a vague predicate and outline the various philosophical accounts - such as the epistemic view, degree theory and supervaluationism - which attempt to address vagueness. I then discuss various well-known approaches to poverty measurement in the light of these accounts. Amongst a range of issues which arise through relating philosophical accounts to poverty measurement are issues relating to multi-dimensionality. I suggest that these have not as yet been adequately addressed by those involved in applying fuzzy poverty measures.

The Chapter is structured as follows. The chief characteristics of vague predicates are described and it is argued that the predicate "poor" has all these characteristics in Sect. 1.2. In Sect. 1.3, the various philosophical accounts of vague predicates are explained and their strengths and weak-

[1] Acknowledgements: I owe a debt of gratitude to those who have inspired my work in this area, and to those with whom I have discussed some of the issues in this Chapter. I thank John Broome, Enrica Chiappero-Martinetti and Amartya Sen for their inspiration and Enrica Chiappero-Martinetti, David Clark and Pasquale Valentini for helpful discussions relating to the arguments in this Chapter.

nesses are evaluated. In Sects. 1.4 and 1.5 these accounts are related to issues in the measurement of poverty, focussing on the epistemic view and fuzzy poverty measures in Sect. 1.4 and supervaluationism and poverty measurement in Sect. 1.5. Sect. 1.6 concludes.

1.2 The Vagueness of "Poor"

A number of predicates[2] are usually classified as vague. Examples include "tall", "bald" and "nice". Furthermore, while the philosophical literature has tended to focus on the predicate vagueness, adverbs (such as "quickly") and quantifiers (such as "many") can also be vague (Keefe and Smith 1996, p 5). I shall focus in this Chapter on the predicate vagueness, since my central claim is that "poor" is a vague predicate. However, "poor" is not the only predicate relevant to poverty measurement. I shall also rely on claims about the vagueness of the predicates "extreme" and "chronic" in addressing issues relating to the measurement of extreme (or "ultra") and chronic poverty. The three well-known and inter-related characteristics of vague predicates are: (1) that they allow for borderline cases, where it is not clear whether the predicate applies or not; (2) that there is no sharp borderline between cases where the predicate does, and does not, apply; and (3) they are susceptible to a Sorites paradox.

First, consider the existence of borderline cases. In the case of the predicate "tall", there are certainly cases where one would unhesitatingly classify people as "tall" and "short". However, there may also be cases where we cannot say that someone is *definitely* tall. One might say, in a case like this, that the relevant person is "borderline" tall. Similarly there are borderline cases of "bald" and "nice". Furthermore, in our ordinary use of "tall", there is no sharp borderline between those who are, and are not, tall. That is, there is no exact height *h* such that anyone who is shorter than *h* is not tall, while everyone else is tall. Similarly there is no exact borderline in the case of "bald". There is no precise number of hairs such that if one has less than that number of hairs on top of one's head one qualifies as bald.

Finally, consider the well-known Sorites paradox or "paradox of the heap". Suppose that John is a tall man. It seems plausible that making John a tiny bit shorter will leave him tall. This suggests that whenever we make a tall man a tiny bit shorter he must still be tall. So if we repeatedly make John a tiny bit shorter, he should remain tall. Yet obviously if we make John a tiny bit shorter enough times he will be short. We are thus led to

[2] A predicate is whatever is affirmed or denied of a subject by means of the copula (i.e. the verb "be") e.g. "mortal" in "all men are mortal".

contradiction - since John will be both tall and short. This is the paradox. Another version of this paradox focuses on a heap. Suppose we are confronted with a heap of pebbles. It seems plausible that taking a single pebble away from the heap will leave a heap of pebbles. One can apply this logic repeatedly each time a pebble is removed, so that each time a pebble is removed we should still be left with a heap of pebbles. Yet if enough pebbles are removed, one by one, all that will be left is a single pebble. One pebble does not make a heap, and so we have a contradiction.

The predicate "poor" has all three characteristics of vague predicates. Consider someone who is poor in terms of income. It seems implausible that giving this person a single penny will make her non-poor. Yet if one repeatedly gives her enough pennies, one by one, she will be rich as regards income. Similarly, there are cases where, in our ordinary usage, we might want to classify someone as "borderline poor". Furthermore, in spite of the use of exact poverty lines to separate out the poor and the non-poor in some official contexts, there does not seem to be a sharp borderline between the poor and the non-poor. So "poor" has all the standard characteristics of vague predicates. It is one among many vague predicates which are used to describe phenomena that social scientists study. It is easy to check that other predicates relevant to poverty measurement - notably the predicates "extreme" and "chronic" - also have the characteristics of vague predicates.

So far, the examples I have focussed on have typically involved only one dimension which is relevant to judging whether or not some predicate applies. In the case of "tall", height was the only relevant dimension. In the case of baldness, the only consideration I invoked was the number of hairs on the top of a person's head. In the case of income poverty, the only dimension was income measured by the number of pennies one had. Yet in the cases of some vague predicates, multiple dimensions are relevant to whether or not the predicate applies. This is the case, for example, with the predicate "nice". Suppose that Jane is extremely polite, sociable and generous, yet is also sometimes bad-tempered. One might say that Jane is not *definitely* nice but only borderline nice. The fact that there are a number of dimensions that are relevant to judging whether Jane is nice is relevant to the fact that she is classified as borderline nice.

This point about multi-dimensionality and vagueness is relevant to the predicate "poor". So far, I have focussed on income poverty. Whether a person was judged to be "poor" was just a matter of the number of pennies, or the amount of income she had. Now consider an alternative example where we allow for multiple dimensions. Suppose that a person can be poor in terms of income, health and educational achievement. Jim, it turns out, has a decent income, but cannot read and write and has a debilitating disease. He classifies as rich in terms of income, but as poor in terms of health and education, even when one allows for vagueness in these dimensions. Or suppose that Jim has a very low income but enjoys good health and is well educated, so that he counts as being poor with regard to income but is not poor in terms of health and education (again when one allows for the vagueness of "poor" in each of these dimensions). Should we classify Jim as poor, all things considered, in these two scenarios? As long as one adopts a multi-dimensional view of poverty, it may not be obvious whether we should judge him to be poor or not poor in either case. Jim may be classified as "borderline poor" in both cases even though the vagueness in this case does *not* relate to the question of whether or not he would qualify as poor in each of the relevant dimensions. The multi-dimensionality of poverty is thus relevant to its vagueness.

Finally, it is worth noting that where a predicate is vague, it is usually argued that there is not merely vagueness about whether or not the predicate applies. If this was all that mattered, it might be that there are three sharply delineated sorts of cases: those where the predicate applies; those where it does not apply; and "indefinite" cases in between. Yet in the case of vague predicates it seems that there are no precise borderlines between these sorts of cases. For example, in the case of "tall" there is no sharp boundary between those cases which are definitely tall and those which are borderline tall. Vagueness about *this* boundary is vagueness about the limits of a rough borderline. It is a form or "vagueness about vagueness" or "higher-order" vagueness.

1.3 Three Views of Vagueness

Philosophical accounts of vagueness typically attempt to address the characteristics of vague predicates (i.e. the existence of borderline cases, rough borderlines and susceptibility to a Sorites paradox) while also allowing for "higher-order" vagueness. Epistemic views of vagueness are distinct because they suppose that even in the case of vague predicates there is actually a precise borderline between cases where the predicate does, and does not, apply. According to such views it is impossible to know where the exact borderline lies. Williamson (1992, 1994) has championed a version of this view. According to Williamson because it is impossible to know the borderline between cases where a vague predicate such as "poor" does, and does not apply, we must leave a "margin of error" in applying the predicate. Inasmuch as there is any vagueness about where the exact borderline between the poor and the non-poor lies, it is just a matter of *ignorance*.

There are a number of problems with this view of vagueness, some of which I shall list here. Firstly, this account simply denies one of the key characteristics of vague predicates which a philosophical account of vagueness needs to address: the non-existence of a precise borderline between cases where the predicate does, and does not, apply. Secondly, in many cases of vagueness it simply seems implausible to suppose that where there are borderline cases, ignorance is the root of the problem. Consider the predicates "tall" and "bald", and borderline cases of "tall" and "bald". It seems very implausible that our considering some particular cases to be "borderline tall" or "borderline bald" reflects any sort of ignorance. There seems to be no knowledge which, if we had access to it, would resolve the question of where the exact borderline is. Furthermore, if someone holding the epistemic view responds by stating that it is impossible to know the exact location of the borderline, that does not in itself help much. What one needs is an account of *why* it is impossible to know the exact location of this borderline. The significant advantage of the epistemic view, for those who hold it, is that it retains classical logic. It retains the "law of excluded middle" according to which, for any predicate such as "tall", either x is, or is not, tall. The vagueness of "tall" does seem to violate this law, since in borderline cases of "tall" someone is neither definitely tall nor definitely not tall. From an epistemic view this is not so: if there *appears* to be any vagueness about whether or not someone is tall, this is just a matter of ignorance. Epistemic views also retain two truth values - "true" and "false" - so that the "principle of bivalence" (which only allows for "true" or "false" statements) holds. Finally, the epistemic view can address the Sorites paradoxes that arise in the case of vague predicates. In the case of "tall", for example, the epistemic view would simply reject the idea that it is always true that slightly reducing the height of a tall man will leave him tall.

The two other well-known accounts of vagueness - degree theory and supervaluationism - drop classical logic. Degree theories do so by supposing that there are more than two truth values. Truth, on these views, comes in degrees. Thus, in borderline cases of "tall", it is true to some degree that people are tall. There are many different forms of degree theory. One sort just adds another value - such as "indefinite" - which holds for cases which fall between those which are definitely true and those that are definitely false. Fuzzy set logic goes further and quantifies the degree of truth in borderline cases. Views of this sort have been illustrated by Zadeh (1965, 1975), Goguen (1969) and Machina (1976). They typically measure a degree of truth on the [0,1] interval, with 0 signifying definite falsehood and

1 signifying definite truth. Unlike epistemic views, degree theories allow explicitly for rough borderlines as well as for the existence of borderline cases. They can also address Sorites paradoxes. If, for example, it is taken to be *nearly* true, or true to a high degree, that each time one takes a penny from a rich person, she remains rich, then repeatedly taking a penny away from a rich person enough times may make it definitely *false* that she is rich. These are clearly strengths of degree theory.

However, degree theory also faces potential criticism. One worry relates to the very idea of a degree of truth. How is one to make sense of this? One way of doing so goes like this. Suppose we are concerned with whether or not x is F-er than y, where F is a vague predicate. Then it might be argued that it is *more true* that x is F than that y is F. Is this plausible? There are cases where it clearly is not at all plausible. Suppose that we are concerned with "tall", and that both John and Jim are very tall. As it happens John is a little taller than Jim. If the account of a degree of truth just given is correct, then it follows that it is more true that John is tall than that Jim is tall. However, since both John and Jim are very tall, this is surely not so: it is simply true that John and Jim are tall. In responding to this point, a degree theorist might argue that the intuition about degrees of truth only applies to borderline cases. Yet there seems to be no reason for which the intuition which underlies this account of degrees of truth should apply only to borderline cases. This seems to be a weakness in degree theories which make sense of degrees of truth in this way.

Some criticisms of degree theory apply specifically to those variations of it - such as fuzzy set theory - which attempt to put a numerical value on the degree of truth. Some worry that it is inappropriate to put numerical values on degrees of truth, because of considerations relating to higher-order vagueness. Assigning a precise numerical value to the degree of truth of vague statements seems inconsistent with allowing for vagueness about the degree of truth. Another related worry concerning these forms of degree theory is that they assume that there is a precise cut-off between those cases that are definitely true or false (i.e. true to degree 0 or degree 1) and those that are not. Proponents of these forms of degree theory can respond in a number of ways. They may respond by suggesting that degrees of truth of statements involving vague predicates are also true to some degree. There can be, on this response, a degree of truth about the degree of truth of a statement. Yet this response may fail to convince many because of the precision involved in assigning numerical values to degrees of truth. Alternatively, a degree theorist may suggest that there *is* a precise cut-off between cases which are definitely true and those which are not, and that higher-order vagueness is just ignorance about the exact degree of truth assigned to a statement. Here degree theorists end up taking a line which is similar to that taken in epistemic views. Again this response might not adequately address the worry about higher-order vagueness if the nature and source of ignorance is not clarified.

Finally, Keefe (1998, 2000) has recently discussed a number of potential problems with degree theories which use numerical values to capture degrees of truth. For measurement of degrees of truth to be possible, a basic requirement is that such degrees can be ordered. For this requirement to hold, it must be the case that all sentences are comparable as regards degrees of truth. Writing "$\exists_T$" for "true to a greater or the same degree", then for all sentences p and q, it must be true that $p \exists_T q$ or $q \exists_T p$. Yet is it obvious that this is so? In the case of predicates involving more than one dimension it is not at all clear that it is. If there are borderline cases of "nice" involving people who are pleasant and unpleasant in quite different ways, then it is not clear that we can compare the degree to which it is true that they are nice in different dimensions. Similarly, it may be difficult to compare the degree of truth of "John is nice" and "the chair is red". Keefe argues that the confusion in attempts to measure degrees of truth might arise from the fact that some vague predicates - such as "tall" - allow for measurement. Yet it is a mistake to jump from the plausible thought that height can be measured to the view that the degree of truth of "Jack is tall" can also be measured, when Jack is borderline tall[3]. The jump might be plausible if the degree of truth of "x is F" is closely related to how $F\,x$ is relative to others - for example whether x is more or less F than y. Yet as we saw earlier, one might resist the claim that "x is F-er than y" implies that "it is more true that x is F than that y is F". Finally, it is also worth mentioning that further complications might arise because it may not be entirely clear which dimensions are relevant to judging whether or not some predicate (such as "nice") actually applies. These various worries about some versions of degree theory in the context of multi-dimensionality are clearly also relevant to the case of poverty. In spite of these potential problems with degree theory, fuzzy set theory is the most widely applied account of vagueness. The fact that it involves numerical values no doubt makes it attractive to economists and social scientists.

An account of vagueness which has not been widely explored by economists and social scientists is supervaluationism. Supervaluationism develops the thought that statements involving vague predicates might, or might not, be true depending on the manner in which they are made more

[3] Smith (2003) argues that Keefe's argument here only applies to some (confused) versions of fuzzy set theory. However, as Keefe (2003) writes in response to Smith, this point does not undermine the claims she makes against degree theories, like fuzzy set theory, which use numbers to capture degrees of truth.

precise. If a statement involving a vague predicate is true in all acceptable ways in which it can be made more precise, one might say that it is "super-true". This is the central intuition running through the best known version of supervaluationism, which has been developed by Fine (1975). On Fine's account, for any vague predicate, there are a number of "admissible" ways of making statements involving the predicate more precise or "precisifying" it. Fine "maps" the various ways of making statements involving a vague predicate more precise in terms of a "specification space". Points in this space include "base points" where the statement is initially specified. These points are "extended" by making the statement more precise. If a statement has not been completely precisified a "partial" specification point has been reached. Once a statement has been made as precise as possible a "complete" specification point has been reached. A vague statement is then "super- true" in Fine's formal sense if and only if it is true in all admissible ways of making it more precise or, equivalently, in all admissible "precisifications".

This account has some attractive features. It clearly allows for borderline cases. From a supervaluationist view these are statements which are true in some, but not all, admissible precisifications. Supervaluationism also allows for rough borderlines since there are a number of admissible ways of drawing borderlines in the case of vague predicates such as "tall" and no single borderline is privileged. Furthermore, this account seems to get round the Sorites paradox. To see why, let's consider "tall". For each admissible way of making a statement such as "John is tall" completely precise there is an exact height *h* such that making John a tiny bit shorter than *h* means that he is not tall. Nonetheless, since no such exact height *h* is privileged, there is no *h* such that (it is super-true that) someone at or above this height is tall, while anyone shorter than *h* is not tall. So, unlike the epistemic view, supervaluationism addresses the Sorites paradox without giving up on the existence of rough borderlines. Finally, Fine's supervaluationism attempts to allow for higher-order vagueness by suggesting that the predicate "admissible" is vague, so that the set of admissible precisifications of a statement is also vague.

Fine has been criticised because his account clearly makes a great deal of use of the notion of precision. To this degree, his approach can be seen as an attempt to address vagueness by insisting on precision. This can be seen as an inappropriate response to vagueness[4]. Finally, it is sometimes argued in defence of supervaluationism that it comes close to preserving classical logic. Firstly, supervaluationism retains the law of excluded mid-

[4] There are also other more technical objections to supervaluationism. See Williamson (1994).

dle in virtue of the fact that for any vague predicate F either x is or is not F for *all* ways of making the borderline between those objects which are or are not F as precise as possible[5]. Finally, unlike degree theory, Fine's version of supervaluationism does not require degrees of truth. This might be an attraction for some. However, some versions of supervaluationism do involve degrees of truth (Lewis 1970; Kamp 1975). Intuitively one statement might be truer than another if it is true on a larger number of admissible precisifications than the other. This point serves to remind us that the three-fold distinction between epistemic views, degree theories and supervaluationism itself has rough borderlines.

1.4 Epistemic and Fuzzy Set Theoretic Views and the Measurement of Poverty

It is important to recognise that while it is only recently that poverty researchers have explicitly begun to take on board the implications of vagueness, the issue has implicitly been addressed in some literature. Most notably in the "mainstream" literature on poverty - which does not explicitly address vagueness - it has been recognised that even if there is an exact cut-off between the poor and the non-poor there may be difficulties about establishing where this cut-off lies. In some of the literature, the problem with establishing an exact cut-off is seen as deriving from the "noisiness" of data on living standards (Ravallion 1994). In the light of such noisy data, there is an advantage in allowing for a range of poverty lines, to allow for a margin of error in making poverty judgements. Clearly, the implicit view of vagueness adopted here is an epistemic one. Furthermore, this approach has the standard problems of an epistemic approach. It seems implausible that even if we had perfect, "noiseless" data we could establish an exact, non-arbitrary, cut-off. It is important to distinguish the issue of noisy data, or ignorance which derives from other sources, from evaluative disagreement. Sometimes it is argued that people differ about where they might set the poverty line because of evaluative disagreement. Given the variety of evaluative judgements, some advocate allowing for a range of poverty lines in making poverty judgements (most notably Atkinson 1987; Foster and Shorrocks 1988). This well-known approach does not address vagueness, though it is quite possible that evaluative judgements can also

[5] On the other hand, supervaluationism violates the principle of bivalence because according to supervaluationism it is *not* the case that all statements are true or false. In cases where statements are true on some, but not all, precisifications, they are neither true or false (Keefe and Smith 1996, p 7).

be imprecise, especially where there are multiple dimensions involved in making such judgements[6]. So vagueness may also be relevant here.

The suggestion that fuzzy set theory might be applied to the economics of poverty and inequality can be traced to Sen's writings. In his writings on economic inequality Sen (1971, p 5) recognised that the notion of inequality "that we carry in our mind is, in fact, much less precise" than that involved in most inequality measures. Sen thought that imprecision implied that inequality rankings are "incomplete" - so that there are cases where, of two states of affairs, it is neither true that one is more unequal than another, nor true that they are equally unequal. He made similar observations in the context of poverty. For example, in his *Poverty and Famines* he wrote that "while the concept of a nutritional requirement is a rather loose one, there is no reason to suppose that the concept of poverty is clear cut or sharp ... a certain amount of vagueness is implicit in both concepts" (Sen 1981, p 13).

In discussing inequality measurement, Basu (1987) argued that Sen had taken an "all or nothing" view in suggesting that we should deal with imprecision by adopting incompleteness, which allows for cases where one cannot make any judgement at all. He argued there are cases that fall between those where one can make a precise judgement and those where one could make no judgement at all: cases where one can only make an *imprecise* judgement. On this basis, Basu developed his axiomatic fuzzy set theoretic measure of inequality. Sen's writings are also supportive of the use of fuzzy set theory and measures based on it. In fact the precision of such measures is clearly an attraction for Sen. If the relevant concept is ambiguous, Sen suggests that "the demands of precise measurement call for *capturing* that ambiguity rather than replacing it with some different idea - precise in form but imprecise in representing what is to be represented" (Sen 1989, p 317). In this context, Sen suggests that fuzzy set theoretic measures and incomplete orderings have quite a bit to offer economics. It is worth noting that it is just this precision with which fuzzy set theoretic accounts of vagueness capture imprecision or ambiguity that worries those who are concerned about higher-order vagueness.

The use of fuzzy set theory unsurprisingly spread to the measurement of poverty with important early contributions from Cerioli and Zani (1990) and Cheli and Lemmi (1995). In the application of fuzzy set theory to poverty measurement, there is typically taken to be a degree to which someone

[6] Literature on fuzzy preferences has emerged in economics to address such evaluative vagueness. See Barrett and Pattanaik (1989) for an introductory survey. There is also literature on vagueness and topics in welfare economics. On this see Broome (2004) and Qizilbash (2005a, 2005b).

is a member of the set of the poor. In the terms used above, the degree of membership captures the degree to which it is true that someone (or some household) belongs to the set of the poor. The membership function to the set of the poor is typically taken to lie on the [0,1] interval, with 0 meaning definite non-membership, 1 definite membership and numbers in between capturing the degree of membership. One key element in this context is the "membership function" which maps an individual's (or household's) performance in terms of an indicator, or in terms of a set of indicators, on to a degree of membership of the set of the poor. The first attempt at fuzzy poverty measurement - advanced by Cerioli and Zani - involved a linear membership function. The simplest version of their measure was income based, though Cerioli and Zani also developed variations on their measure which allow for the multi-dimensionality of poverty. In the simplest case, they took a level of income at or below which a person (or household) is judged to be definitely (income) poor, and one at or above which she (or it) is taken to be definitely not (income) poor. In between these levels, the degree of membership of the set of the poor increases in a linear way as income falls. In the multi-dimensional context, Cerioli and Zani suggested alternative measures, based on the same approach. In one variation, they suggested an ordinal ranking of levels of disadvantage for each dimension. In each dimension there is some level at or below which a person (or household) counts as definitely poor, and one at or above which she (or it) classifies as definitely not poor. In between these levels, the degree of membership of the set of the poor (for each dimension) depends on the person's (household's) position in the ordinal ranking. Cerioli and Zani explored various ways of weighting the dimensions of poverty in judging whether or not a person (or household) is definitely poor taking into account all the dimensions of poverty. It is worth noting that the various approaches they discuss imply that - as long as each dimension has positive weight - a person (or household) must qualify as definitely poor on all dimensions - i.e. get a score of 1 on all dimensions - to gain a score of 1 overall and to count as definitely poor overall.

In their important contribution, Cheli and Lemmi (1995) criticised the arbitrary use of two critical levels which define the range of levels of income or other indicators where there is fuzziness in Cerioli and Zani's methodology. They suggested an alternative "Totally Fuzzy and Relative" (TFR) approach. The approach works so that the cut-offs used to establish the relevant range of levels is driven by the distribution itself. The TFR approach can be applied to both income and multi-dimensional contexts. Only those who are most (least) deprived in terms of the distribution of the relevant indicator (which may be income or some indicator used in a multi-dimensional application) are definitely poor (not poor) in terms of that indicator. Between these levels, the degree of membership of the set of the poor in terms of the relevant indicator depends on, and "mirrors" the distribution of the relevant indicator. Like Cerioli and Zani, Cheli and Lemmi suggest a multidimensional variation of their measure which involves weighting. The TFR approach has now been applied in a number of contexts, and Chiappero-Martinetti (1994, 1996, 2000) has made a number of influential applications in the Italian context.

Some issues that arise in the context of applications of these measures relate directly to points raised in the context of the degree theories of vagueness discussed earlier. Firstly, just as there were problems with providing an account of degrees of truth, there are problems with giving an intuitive interpretation of a measure of the degree of membership of the set of the poor. Secondly, the issue of comparability of degrees of truth in different dimensions also arises for multi-dimensional poverty measures. Are we right to assume that we can compare degrees of membership in diverse dimensions such as health, education and housing? If we cannot, it is perhaps best to use fuzzy measures in specific dimensions without attempting to form judgements across dimensions. Finally, again relating to the issue of dimensions, the Cerioli and Zani and TFR approaches take the dimensions of poverty as given. As we saw in the discussion of vague predicates, this may not be sensible: we may not be able to pin down precisely the range of dimensions which are relevant to poverty measurement.

1.5 Supervaluationism and the Measurement of Poverty

In an earlier paper (Qizilbash 2003) I attempted to address some of the problems that arise for fuzzy poverty measures by developing a framework which is inspired by supervaluationism. An intuitive interpretation of fuzzy poverty measures emerges in this framework. I only sketch the framework in broad terms here so as to show how it attempts to address some of the problems just noted. First of all, I follow Kit Fine in allowing for a set of admissible specifications of "poor". The set of such specifications can, of course, be vague (because of the vagueness of "admissible"). Each admissible specification involves a set of dimensions of poverty and a range of critical levels relating to each dimension. Any dimension of poverty which appears on *all* admissible specifications is termed a *core* dimension. In each dimension, someone (or some household) who falls at or below the lowest admissible critical level is judged to be *definitely* poor in that dimension. If she (it) is definitely poor on a core dimension, she (it) is *core poor*. Someone (or some household) who falls at or above the highest critical level is definitely not poor in that dimension. Anyone (or any household) who (that) is definitely not poor on all admissible dimensions is *non-poor*. Those who are neither core poor nor non-poor fall at the *margins of poverty*. If someone is core poor, I have suggested that it is "super-true" that he/she is poor - in Fine's terms - since he/she falls at or below the lowest critical level in a dimension which is admissible on *all* specifications of "poor". There is no ambiguity about whether or not such a person is poor, taking account of all the dimensions of poverty. So, for example, if nutrition is a core dimension and someone falls at or below the lowest admissible critical level one would classify that person as core poor without worrying about how he/she is doing on other dimensions.

In this framework, fuzzy poverty measures can be interpreted as measures of "vulnerability" in each dimension. In each dimension, there will be some who falls between the highest and lowest critical levels, and so are neither definitely poor nor definitely not poor in that dimension. These people (or households) can be seen as "vulnerable" in as much as they are poor in terms of *some* admissible critical level in the relevant dimension, and would be defined as poor if that critical level was used. Fuzzy poverty measures capture how "close" these individuals (or households) come to being definitely poor in the relevant dimension. This, intuitively, is the sense of vulnerability which is relevant to the interpretation of fuzzy poverty measures. On this interpretation, the Cerioli and Zani measure is a linear measure of vulnerability while Cheli and Lemmi provide a relative measure. However, it is worth being clear about what is meant by "vulnberability" here, given the way in which Cheli and Lemmi express the intuition behind their measure. They write that the "membership function will express the exposure of risk to poverty" (Cheli and Lemmi 1995, p 129). There is scope for confusion here because much of the discussion of "vulnerability" in economics and development studies has to do with the risk of *becoming* poor as a consequence of some event. That sense of vulnerability also clearly relates to the "exposure of risk to poverty" and focuses on the *probability* of some person (or household) falling below some (possibly exact) borderline (see, for example, Morduch 1994).

The notion of vulnerability which underlies the interpretation of fuzzy poverty measures in my framework is different. Fuzzy measures are conceived as measures on the "specification space" (in Fine's terms) in a particular dimension. So they relate to the range of precisifications of "poor" on which someone is judged to be poor in a particular dimension. As the range or proportion of precisifications on which someone classifies as poor in a particular dimension increases that person classifies as more vulnerable[7]. In this context, anyone who is defined as poor on all but one critical level (or a very small proportion of critical levels) in some dimension might classify as "extremely vulnerable", in the sense that a tiny relaxation of the standards used for judging whether or not someone is definitely poor in that dimension will lead to that person classifying as definitely poor[8]. So if one uses Cheli and Lemmi's notion of "exposure to risk of poverty" in my framework, this must be interpreted in terms of the notion of vulnerability described here. Obviously whether or not someone counts as vulnerable in the framework *might* be related to whether or not she is vulnerable in the "standard" sense. However, the two senses of vulnerability are quite distinct. In using the term "vulnerability" to capture the intuition underlying fuzzy measures when they are interpreted within this framework, there is obviously a danger of confusion for those who use alternative notions of vulnerability. Nonetheless, the notion of vulnerability involved in the interpretation of fuzzy measures enriches the analysis of vulnerability by introducing a new conception of it. There is no reason why the most commonly used interpretation of vulnerability in economics should be the only one that is permitted. Finally, it is worth noting that the interpretation of fuzzy poverty measures within this framework is related to the interpretation of degrees of truth on those versions of supervaluationism which overlap with degree theory (Lewis 1970; Kamp 1975). In those versions, the possibility of some form of measurement on the specification space is the underlying intuition for degrees of truth. Yet one need not accept degrees of truth to accept the interpretation of fuzzy measures as measures of vulnerability.

One advantage of the framework sketched here is that it allows for two kinds of vagueness. It allows for vagueness about the critical level at or below which a person (household) classifies as poor. This is "vertical vagueness". It is the focus in the literature on fuzzy poverty measures. However, my framework also allows for vagueness about the dimensions of poverty. As we saw, Keefe (1998) raised this issue in the context of her critique of accounts of vagueness which use numerical values. In the case of a predicate like "nice", the set of dimensions which is relevant to applying the predicate is not sharply defined. This is also true of the predicate "poor".

---

[7] One difference between the Cerioli and Zani and Cheli and Lemmi measures, when they are interpreted in this way, has to do with the way in which the specification space is defined. In the Cerioli and Zani measure it merely has to do with the range of critical levels, while in the TFR methodology it is driven by the distribution. On this see Qizilbash (2003).

[8] This sense of "extremely vulnerable" is used in Qizilbash (2002). It is worth noting that the vagueness of "extremely" would be relevant if one were to develop this idea much further within this framework.

In my framework such vagueness about the dimensions of poverty is "horizontal vagueness". The distinction between dimensions which are core and other admissible dimensions of poverty reflects such vagueness. Chiappero-Martinetti (2005) has cast some doubt on the notion of horizontal vagueness, suggesting that the underlying issue here may be the "complexity" of poverty, which is, in part, constituted by its multi-dimensionality. Chiappero-Martinetti here tries to distinguish issues relating to multi-dimensionality and vagueness in a sharp way. Yet we saw earlier that multi-dimensionality is often invoked in the context of the vagueness of some predicates. Furthermore, in some accounts of poverty, horizontal vagueness can be motivated by the use of the predicate "basic", when poverty is seen in terms of falling short of some "basic" standard. In versions of the "basic needs" approach (Streeten et al. 1981) and in Sen's capability approach - which involves the notion of "basic capability failure" (Sen 1992, 1999) - researchers need to decide on those dimensions of well-being, or those capabilities, that count as "basic". Yet it is highly plausible that "basic" is a vague predicate. Certainly there seems to be no sharp borderline between those needs or capabilities which are, and are not, "basic"[9]. So in these accounts, horizontal vagueness might relate to what is, and is not, judged to be "basic". Neither the Cerioli and Zani measure, nor the TFR methodology (nor any other poverty measure I know of) accommodates such vagueness.

There is a number of further issues about the use of fuzzy poverty measures and the framework sketched here which are worth noting. First, vertical vagueness is often confused with the depth of poverty. Indeed fuzzy measures of poverty are sometimes confused with measures of the depth of poverty. It should be clear that this is a mistake. For any measure of the depth of poverty, we need to establish some critical level relative to which one might measure how far someone who is judged to be poor falls. Measures of the depth of poverty thus usually begin with some precise poverty cut-off and "resolve" vertical vagueness in some arbitrary way. In this context, the vagueness of "extreme" is also relevant. Since whether or not a poor person's (household's) condition is judged to be extreme is the key to whether or not that person (household) is treated as "extremely poor" or "ultra poor", the adverb "extremely" in "extremely poor" is also no doubt vague. Certainly, there appears to be no exact borderline between those who are, and are not, extremely poor. So over and above any vagueness about whether or not someone (or some household) classifies as poor, there is further vagueness about whether that person (or household) quali-

[9] Indeed, this would be true even in the absence of evaluative disagreements about what counts as a "basic" need or capability.

fies as extremely poor. The framework sketched above can easily allow for this further level of vagueness. It would do so by adding a set of admissible critical levels for someone to qualify as "extremely" poor in each dimension. If someone (some household) fell at or below the lowest of these, she (it) would be definitely extremely poor in that dimension. If she (it) was definitely extremely poor in a core dimension, one might say that she is "extremely core poor". However, the use of this term would be misleading if it were taken to imply that core poverty comes in degrees. It does not.

The same basic point holds in the case of the width of poverty. The number of dimensions on which someone (or some household) is poor is the central focus when measuring the width of poverty. Yet measuring width is quite different from capturing horizontal vagueness. Only when issues of horizontal vagueness are resolved - so that the dimensions of poverty are clearly defined - is it possible to measure the width of poverty. Again if one wanted to examine whether or not a person (household) is extremely poor - as regards the width of poverty - one would need to allow for the vagueness of "extreme".

In this context, it is also worth mentioning the amount of time someone (or some household) has been poor. In the literature on poverty measurement, the distinction is sometimes made between those who are "temporarily" (or "transitory") poor and those who are "chronically" poor. Here again there is more than one level of vagueness. On the one hand, there is vagueness about whether or not someone (some household) is poor at a point in time. This is addressed by the framework described above. Once this issue is settled, there is the further issue of whether that person's (household's) condition is "chronic". Since "chronic" is a vague predicate, fixing on any precise number of years (months or other time units) one must be poor to be counted as chronically poor in some dimension is arbitrary, and the vagueness of the predicate "chronic" needs to be taken into account. The framework outlined above can be easily extended to allow for this further level of vagueness, by allowing for a range of admissible periods of time for which a person (household) has been poor in some dimension for that person's (household's) condition to be classified as chronic in that dimension. If a person (household) qualifies as poor for all the relevant admissible time periods for some dimension and critical level, that person's (household's) condition would classify as definitely chronic in terms of the relevant dimension and critical level. If her (its) condition is definitely chronic in a core dimension for the lowest admissible critical level, one might say that she is "chronically core poor"[10]. Again, this term would be misleading if it suggested that core poverty comes in degrees.

Finally, it is worth noting another way in which the framework described here differs from standard approaches which use fuzzy poverty measures in the context of multi-dimensionality. As we saw in the previous section, in most fuzzy set theoretic measures one has to qualify as definitely poor on all dimensions to qualify as definitely poor overall, as long as all dimensions have positive weight in arriving at the overall judgement. By contrast, in the supervaluationist approach outlined here one only needs to be definitely poor on a core dimension to be defined as core poor, so that one is poor on all admissible specifications of "poor". I think this is intuitively forceful, since one might want to classify someone who is starving as unambiguously poor irrespective of how she is doing in terms of other dimensions. Nonetheless, it is worth noting that while this holds on the framework I have developed, it is possible to develop supervaluationism differently. One might, for example, develop it so that it is only "super-true" that someone (some household) is "poor" if she (it) is poor for all admissible dimensions and critical levels. If one developed supervaluationism in this way it would be compatible with the standard form that multi-dimensional fuzzy poverty measures take. However, I would resist this version of supervaluationism. To see why, consider a case where there are just three dimensions of poverty, involving education, health and housing. If one pursued this variation of supervaluationism it would *not* be super-true that a person is poor, even if she is starving and illiterate as long as she happens to live in a high quality house. If find this both implausible and unattractive.

[10] My articulation of this idea emerged through discussion with Clark, who was, at the time, working on extending or modifying the framework to allow for time. Clark first used the term in work in progress co-authored with Hulme (Clark and Hulme 2005). In parts of their text Clark and Hulme use it in the same sense that I am using it here. However, their analysis is distinct, and they propose a notion of "temporal" vagueness, alongside "horizontal" and "vertical" vagueness. One difference between my view and that adopted by Clark and Hulme is that they would not take someone to be unambiguously poor if she were core poor at a point in time. They only classify the chronically core poor as unambiguously poor. They would, thus, not be able to judge that a famine victim who is very seriously malnourished at a point in time is unambiguously poor. I find this highly implausible. By contrast, my view is that one must separately establish whether some person (or household) is core poor at a point in time - this would imply that there is no ambiguity about whether or not the person is poor at that moment - and whether that person's (household's) condition is definitely chronic.

1.6 Conclusions

Vagueness must be addressed by those who attempt to measure poverty because "poor" is a vague predicate. Philosophers have developed a range of different accounts of vagueness, of which degree theory is one. Fuzzy set theory is one particular form of degree theory. Some problems with fuzzy set theory - as a theory of vague predicates - arise from the precision with which it attempts to capture vagueness. Others arise from its attempt to measure degrees of truth when multiple dimensions are involved in the application of a predicate. While the precision with which fuzzy set theory attempts to capture vagueness appears to be a problem when it comes to higher-order vagueness, it is this very precision and the use of numerical values to capture degrees of truth which makes it attractive to some economists. Problems regarding multi-dimensionality arise for fuzzy set theory both as an account of vagueness and as a methodology for measuring poverty. An alternative framework which is inspired by supervaluationism can allow for vagueness about the dimensions of poverty, while also providing a more intuitive interpretation of fuzzy poverty measures. This framework can also be extended to allow for the vagueness of predicates such as "extreme" and "chronic". However, this framework addresses the multidimensionality of poverty in a way which is quite different to that implicit in some fuzzy poverty measures. This multidimensionality will need further attention in future attempts to develop fuzzy poverty measures.

References

Atkinson AB (1987) On the measurement of poverty. Econometrica 55:749-764

Barrett CR, Pattanaik PK (1989) Fuzzy sets, preference and choice: some conceptual issues. Bulletin of Economic Research 41:229-253

Basu K (1987) Axioms for a fuzzy measure of inequality. Mathematical Social Science 14:275-288

Broome J (2004) Weighing Lives. Oxford University Press, Oxford

Cerioli A, Zani S (1990) A fuzzy approach to the measurement of poverty. In: Dagum C, Zenga M (eds) Income and Wealth Distribution, Inequality and Poverty. Springer Verlag, Berlin, pp 272-284

Cheli B, Lemmi A (1995) A "Totally" Fuzzy and Relative Approach to the Multidimensional Analysis of Poverty. Economic Notes 24:115–134

Chiappero-Martinetti E (1994) A new approach to the evaluation of well-being and poverty by fuzzy set theory. Giornale Degli Economisti e Annali di Economia 53:367-388

Chiappero-Martinetti E (1996) Standard of living evaluation based on Sen's approach: some methodological considerations. Notizie di Politeia 12:37-53

Chiappero-Martinetti E (2000) A multi-dimensional assessment of well-being based on Sen's functioning theory. Rivista Internazionale di Scienze Sociali 108:207-231

Chiapperio-Martinetti E (2005) Complexity and vagueness in the capability approach: strengths or weaknesses? In: Comim F, Qizilbash M, Alkire S (eds) The capability approach: concepts, applications and measures. Cambridge University Press, Cambridge: forthcoming

Clark D, Hulme D (2005) Towards a unified framework for understanding the depth, breadth and duration of poverty. Typescript, University of Manchester

Fine B (1975) Vagueness, truth and logic. Synthese 30:265-300

Foster JE and Shorrocks AF (1988) Poverty orderings and welfare dominance. Social Choice and Welfare 5:91-110

Goguen JA (1969) The logic of inexact concepts. Sythese 19:325-375

Kamp JAW (1975) Two theories of adjectives. In: Keenan EL (ed) Formal semantics of natural language Cambridge University Press, Cambridge, UK, pp 123-155

Keefe R (1998) Vagueness by numbers. Mind 107:565-579

Keefe R (2000) Theories of vagueness. Cambridge University Press, Cambridge, UK

Keefe R (2003) Unsolved problems with numbers: reply to Smith. Mind 112:291-293

Keefe R, Smith P (1996) Vagueness: a reader. MIT press, Cambridge Mass

Lewis D (1970) General semantics. Synthese 22:18-67

Machina K (1976) Truth, belief and vagueness. Journal of Philosophical Logic 33:203-251

Morduch J (1994) Poverty and vulnerability. American Economic Review 84:221-225

Qizilbash M (2002) A note on the measurement of poverty and vulnerability in the South African context. Journal of International Development 14:757-772

Qizilbash M (2003) Vague language and precise measurement: the case of poverty. Journal of Economic Methodology 10:41-58

Qizilbash M (2005a) Transitivity and vagueness. Economics and Philosophy 21:109-131

Qizilbash M (2005b) The mere addition paradox, parity and critical level utilitarianism. Social Choice and Welfare 24:413-431

Sen AK (1973) On economic inequality. Oxford University Press, Oxford

Sen AK (1981) Poverty and famines: an essay on entitlement and deprivation. Clarendon Press, Oxford

Sen AK (1989) Economic methodology: heterogeneity and relevance. Social Research 56:299-330

Sen AK (1992) Inequality reexamined. Oxford University Press, Oxford

Sen AK (1999) Development as freedom. Oxford University Press, Oxford

Smith NJJ (2003) Vagueness by numbers? No worries. Mind 112:283-289

Streeten P, Burki SJ, Haq M, Hicks N, Stewart F (1981) First things first: meeting basic human needs in the developing countries. Oxford University Press, Oxford

Williamson T (1992) Vagueness and ignorance. Proceedings of the Aristotelian Society 66:145-162
Williamson T (1994) Vagueness. Routledge, London
Zadeh LA (1965) Fuzzy sets. Information and Control 8:338-353
Zadeh LA (1975) Fuzzy logic and approximate reasoning. Syntheses 30:407-440

2 The Mathematical Framework of Fuzzy Logic

Bernard Fustier

University of Corsica

2.1 Introduction

In spite of what it may seem, fuzzy logic is not a vague reasoning with indistinct results. On the contrary, it is a rigorous tool that makes it possible for humans to overcome the subtle blend of imprecision and uncertainty of the real world.

It is well-known that fuzzy logic was introduced by Zadeh (1965) in a seminal article entitled “Fuzzy Sets”. This new way of reasoning is based on a very natural principle, the graduality principle, which extends the two-valued classical logic to a more general one where fuzziness is accepted as a matter of science. In particular, we accept that a given proposition is more or less true (or untrue) rather than only true or false. Thus, fuzzy logic can be applied to all these concepts where it is impossible to carried out description in classical mathematical terms because of their natural vagueness.

It seems that the poverty concept falls within the field of fuzzy logic.

However, the majority of applications is still in the industrial world, principally in Japan and Germany (Zimmermann 1993) where fuzzy technology is on the increase with fuzzy tools and fuzzy products such as video cameras, pattern recognition devices etc... Paradoxically, in the area of “soft” sciences, fuzzy logic is of lower penetration. The term of “fuzzy economics” was used for the first time in the summer of 1985 at the First International Fuzzy System Association Congress held at Palma of Mallorca (Ponsard and Fustier 1986). It was the outcome of a long series of research initiated by Ponsard, particularly in the framework of spatial economic analysis (Ponsard 1981a, 1981b, 1982, 1988). Since that time, there has been a certain lack of interest in economic applications of fuzzy subset theory in academic research.

The Chapter is divided into three sections. Sect. 2.1 deals with the graduality principle which applies to “graded” concepts such as fuzzy propositions, fuzzy subsets and fuzzy number concepts. In Sect. 2.2 the basic connectors used in fuzzy logic are illustrated. In Sect. 2.3 the reader can revise the above-mentioned notions referring first to the elaboration of a decision-making process, and then to the construction of a simple model of evaluation.

2.2 The graduality principle

As its name implies, the graduality principle is a principle of graded concepts, a principle in which everything is a matter of degree. In this section, firstly, the fuzzy proposition concept is examined and, then, the fuzzy subset and fuzzy number concepts are examined.

2.2.1 Fuzzy propositions

Let us consider a property p defined on a set X of elements x. We designate by p(x) the degree of truth of the statement "x possesses p" denoted by P(x).

The logic in its classical form recognizes two possibilities (and only two) to express the truth value of any proposition, that is "true" or "false". According to the custom fixed by Boole the truth value is equal to 0 or 1 when the statement is false or true respectively. In other words, p(x) takes its values in the set {0,1} and P(x) is said to be an *ordinary* proposition. This notion supposes that properties p are rigorously defined on the referential sets like, for instance, the masculine gender if we consider a set of persons. In that case, the set {0,1} is enough to express truth values (any intervening state between false and true is excluded). Nevertheless, the two-valued (boolean) logic does not hold out against the pervasive imprecision of the real world. In particular, most properties used in natural languages are rather ill-defined. Thus, to estimate the degrees of truth of statements such as "x is a sympathetic person" or "x is a beautiful woman", it is clear that we need a set of values larger than {0,1}.

Lukasiewicz's (1928) three-valued logic was a first attempt to make the classical logic suppler (the 0.5 value is used when we have doubts about the true value of a proposition). More general logics (multivalued logics) were worked out afterwards, but it is to Zadeh (1965) that we owe the most general one. Indeed, the interval [0,1] substitutes for the set {0,1}. When p(x) belongs to [0,1], P(x) is a *fuzzy* proposition. P(x) is true when p(x) = 1, untrue when p(x) = 0 and "more or less" true (or untrue) for other values of the interval. Notice that [0,1] includes an infinity of values, thus the transition from truth to untruth is gradual rather than abrupt.

The graduality principle applies also to the subset and number notions.

2.2.2 Fuzzy subsets, fuzzy numbers

Let P be a subset of X such that it regroups all the elements x characterized by property p, we can write:

$$P = \{(x / p(x)) \mid x \in X\} \quad (2.1)$$

p(x) is the degree of membership of x to P, that is to say the degree of truth of P(x).

If $p(x) \in \{0,1\}$, then P is an *ordinary* subset of X. If $p(x) \in [0,1]$, then P is a *fuzzy* subset of X. Let us notice that X is an ordinary set, i.e. a non-fuzzy set.

Examples: X = {a,b,c,d} represents a set of regions, if a and d are two islands, b and c two mainland regions, then the ordinary subset of "insular" regions can be written as follows: A = {(a / 1), (b / 0), (c / 0), (d / 1)}. In the classical sets theory it is customary to exclude the elements associated with a zero membership value, one can simply write A = {a,d}. In the case of fuzzy subsets it is not so easy. Because a fuzzy subset is a collection of objects with unsharp boundaries, we have to review each element of X in order to indicate its membership degree. For instance, the "wealthy" regions fuzzy subset of X can be represented by B = {(a / 0.4), (b / 0.8), (c / 0.5), (d / 0.6)}. Let us observe that some membership values can be equal to 0 or/and 1, for instance the fuzzy class of regions with "mild weather" can be represented by the following fuzzy subset: C = {(a / 1), (b / 0.6), (c / 0), (d / 0.8)}. Given P the fuzzy subset defined by (2.1), we give the basic definitions:

- *height* H_P of P :

$$Hp = \vee[p(x)|x \in X] \quad (2.2)$$

where $\vee$ represents the max-operator.

- *kernel* K_P of P:

$$Kp = \{x \in X \quad such \quad that \quad p(x) = 1\} \quad (2.3)$$

- *cardinality* |P| of P:

$$|P| = \sum[p(x)|x \in X] \quad (2.4)$$

Furthermore,

$$\text{P is said } \textit{normalized} \text{ if } H_P = 1 \quad (2.5)$$

and P is *empty*

$$(P = \emptyset) \quad if \quad \forall x \in X : p(x) = 0 \tag{2.6}$$

Remark: in the particular case where each element of X belongs entirely to P, we have K_P = X and |P| = |X|. In other words, P is nothing but the universe X.

Considering C the fuzzy subset of regions with "mild weather", we have: H_C = 1, thus C is normalized. Moreover K_C = {a}and |C| = 2.4, C is non-empty.

In the specific case where X is the set of real numbers (p(x) is a continuous real mapping), it is possible to introduce the convexity notion. For any pair of real numbers x and x', and for any value λ of [0,1], P is said to be *convex* if:

$$p[\lambda x + (1-\lambda)x'] \geq p(x) \wedge p(x') \tag{2.7}$$

where ∧ is the min-operator.

By definition, a *fuzzy number* P is a fuzzy subset of the real line which is normalized and convex such that exactly one real number x_0 exists, called the *mean value* of P, with $p(x_0) = 1$.

When X is a set of discrete values, such as the set of integers, a fuzzy number P can be represented as follows in Figure 2.1.

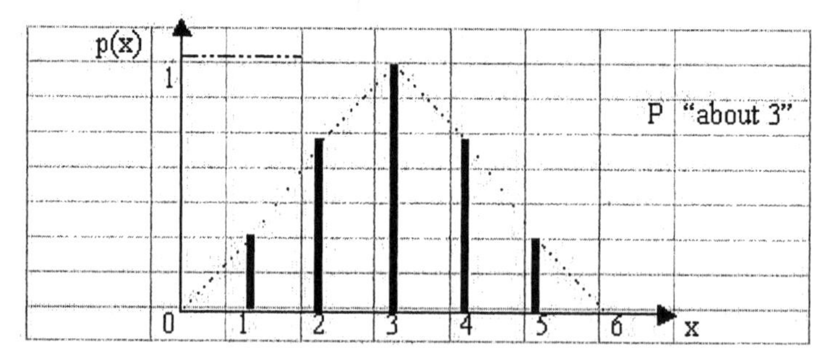

Fig. 2.1. Fuzzy number

On Figure 2.2, the fuzzy subset Q is normalized but not convex: Q is not a fuzzy number, but a *fuzzy quantity*.

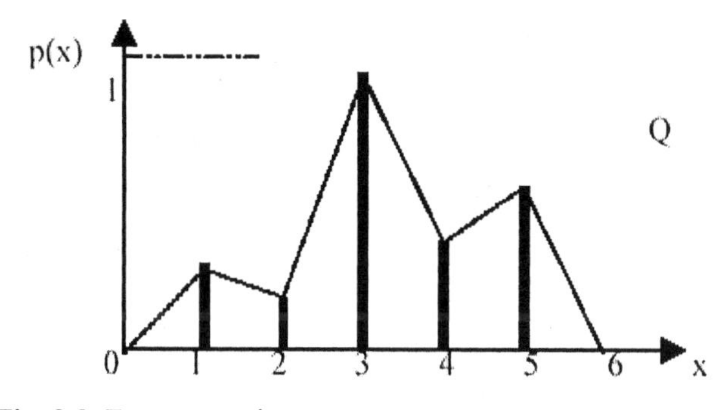

Fig. 2.2. Fuzzy quantity

2.3 The connectors of fuzzy logic

Connectors are operators used to combine fuzzy propositions with the conjunction "and", the disjunction "or", or to express the negation of a given statement.

The min, max operators and the complementation (to 1) were first introduced by Zadeh (1965) to express the "and", the "or" and the "not" respectively.

Other operators have also been suggested. We shall investigate here the basic class of triangular norms and conorms which generalize the use of the min and max operators.

2.3.1 Zadeh's operators

Considering the degrees of truth p(x) and q(x) of the fuzzy propositions P(x) and Q(x), Zadeh's operators are given in Table 2.1.

Table 2.1 Zadeh's operators

proposition:	meaning:	degree of truth:
P(x) and Q(x)	"x possesses p and q"	$p(x) \wedge q(x)$
P(x) or Q(x)	"x possesses p or q"	$p(x) \vee q(x)$
non-P(x)	"x does not possess p"	$1 - p(x)$

$\wedge$ represents the min-operator, $\vee$ represents the max-operator

For example let P(x) be "x is a rich person" with p(x) = 0.6. Assuming that "poor" is the opposite of "rich" in such a way that non-P(x) means "x is a poor person". Under these conditions, the level of truth of:

1. P(x) and non-P(x), i.e. "x is a rich and poor person", is $0.6 \wedge (1 - 0.6) = 0.4$
2. P(x) or non-P(x), i.e. "x is a rich or a poor person", is $0.6 \vee (1 - 0.6) = 0.6$.

Here we want to emphasize that non-contradiction and excluded middle laws no longer exist in the fuzzy logic context.

In the classical logic, the non-contradiction law means that it is impossible to assert an event and its opposite simultaneously, in other words P(x) and non-P(x) is always false. We see here that P(x) and non-P(x) is not untrue, but a slightly true proposition (0.4). Conversely, in the classical logic a proposition such as P(x) or non-P(x) is always true (excluded middle law). In the case of fuzzy logic, a proposition like "x is a rich or a poor person" is not totally true (0.6 instead of 1) because, between being "rich as Croesus" and being "poor as Job", there are still many middle situations that characterize a given person.

Let us note that the $\wedge$ and $\vee$ operators satisfy the following generalization of De Morgan's laws:

$$1-[p(x)\wedge q(x)]=[1-p(x)]\vee[1-q(x)] \tag{2.8 a}$$

$$1-[p(x)\vee q(x)]=[1-p(x)]\wedge[1-q(x)] \tag{2.8.b}$$

The $\wedge$ and $\vee$ operators are said to be *dual* for the complementation. The duality relations (2.8.a) and (2.8.b) are important because they establish a logical link between the $\wedge$ and $\vee$ operators via the complementation. In the following paragraph, we shall see that the complementation to 1 is also used to show the duality between operators different from $\wedge$ and $\vee$. Now, let us consider $P=\{(x/p(x))|x\in X\}$ and $Q=\{(x/q(x))|x\in X\}$ two fuzzy subsets of X. The intersection $\cap$, the union $\cup$ and the complementation * operations correspond to the logical "and", "or" and "not" respectively, thus we have:

$$P\cap Q=\{(x/p(x)\wedge q(x))|x\in X\} \tag{2.9}$$

$$P\cup Q=\{(x/p(x)\vee q(x))\,|\,x\in X\} \tag{2.10}$$

$$P^*=\{(x/1-p(x))\,|\,x\in X\} \tag{2.11}$$

Let us go back with B = {(a / 0.4), (b / 0.8), (c / 0.5), (d / 0.6)} the "wealthy" regions fuzzy subset. We obtain B* = {(a / 0.6), (b / 0.2), (c / 0.5), (d / 0.4)} the "non-wealthy" regions fuzzy subset, $B \cap B^*$ = {(a / 0.4), (b / 0.2), (c / 0.5), (d / 0.4)} the fuzzy subset of regions which are simultaneously "wealthy" and "non-wealthy", and then $B \cup B^*$ = {(a / 0.6), (b / 0.8), (c / 0.5), (d / 0.6)} the fuzzy subset of regions which are either "wealthy" or "non-wealthy". We must observe that:

1. $B \cap B^* \# \varnothing$ (some regions possess both wealth and poverty features)
2. $|B \cup B^*| < 4$ (the union of wealthy and non-wealthy regions does not give the universe, because there are still many middle regions).

Obviously, these definitions apply to the fuzzy numbers.

Example (Zimmermann 1991, p 18): let us consider two fuzzy real numbers P and Q. The meaning of P is "x is considerably larger than 10" with:

$$p(x) = \begin{cases} 0 & \text{if } x \leq 10 \\ \left[1 + (x-10)^{-2}\right]^{-1} & \text{otherwise} \end{cases}$$

the meaning of Q is "x is approximatively equal to 11" with:

$$q(x) = [1 + (x - 11)^4]^{-1}$$

Then the fuzzy number $P \cap Q$ means "x is considerably larger than 10 and approximatively equal to 11". Let us write $f(x) = p(x) \wedge q(x)$, we have:

$$f(x) = \begin{cases} 0 & \text{if } x \leq 10 \\ \left[1 + (x-10)^{-2}\right]^{-1} \wedge \left[1 + (x-11)^{4}\right]^{-1} & \text{otherwise} \end{cases}$$

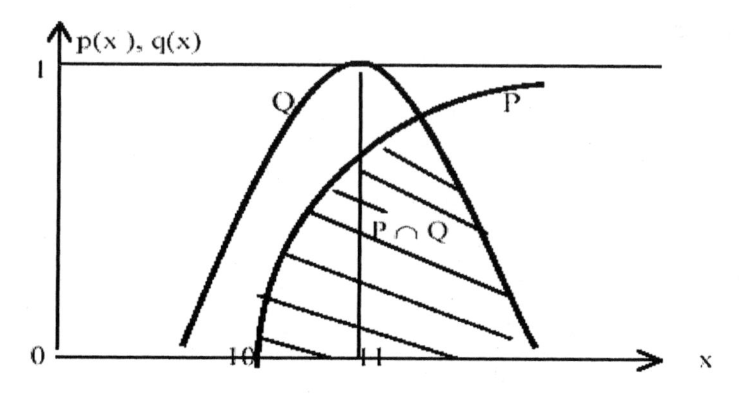

Fig. 2.3. Intersection of fuzzy numbers

The intersection is represented by the curves bordering the hachured part (Figure 2.3).

Algebraic operations with fuzzy numbers have been defined (Dubois and Prade 1979, 1980, 1991; Zimmermann 1991). We consider here the cases of fuzzy addition and fuzzy product.

Let p(x) and q(y) be the membership degrees of the real numbers x and y to fuzzy real numbers P and Q respectively, let P+Q and P•Q be the sum and the product of P and Q respectively. Under these conditions:

1. given the z real values such that z = x + y, where + is the ordinary addition, the membership degree of z to P+Q , denoted f(z), is defined by:

$$f(z) = \vee\, [\, p(x) \wedge q(y) \,|\, z = x + y] \tag{2.12}$$

2. given the z real values such that z = x . y, where . is the ordinary multiplication, the membership degree of z to P•Q, denoted g(z), is defined by:

$$g(z) = \vee\, [\, p(x) \wedge q(y) \,|\, z = x \,.\, y] \tag{2.13}$$

To simplify matters, let us put ourselves in the context of discrete values, for instance P and Q are fuzzy numbers defined on the set of integers such as:

P = {(0 / 0), (1 / 0.5), (2 / 1), (3 / 0.5), (4 / 0)} : "x is approximatively equal to 2"

Q = {(1 / 0), (2 / 0.6), (3 / 1), (4 / 0.6), (5 / 0)} : "y is approximatively equal to 3".

For the addition, the z values are given in the following table:

x \ y	1	2	3	4	5
0	1	2	3	4	5
1	2	3	4	5	6
2	3	4	5	6	7
3	4	5	6	7	8
4	5	6	7	8	9

for z = 1: $f(1) = 0 \wedge 0 = 0$
for z = 2: $f(2) = \vee\, [(0.5 \wedge 0), (0 \wedge 0.6)] = 0$
for z = 3: $f(3) = \vee\, [(1 \wedge 0), (0.5 \wedge 0.6), (0 \wedge 1)] = 0.5$
for z = 4: $f(4) = \vee\, [(0.5 \wedge 0), (1 \wedge 0.6), (0.5 \wedge 1), [(0 \wedge 0.6)] = 0.6$
for z = 5: $f(5) = \vee\, [(0 \wedge 0), (0.5 \wedge 0.6), (1 \wedge 1), (0.5 \wedge 0.6), (0 \wedge 0)] = 1$
for z = 6: $f(6) = \vee\, [(0 \wedge 0.6), (0.5 \wedge 1), (1 \wedge 0.6), [(0.5 \wedge 0)] = 0.6$
for z = 7: $f(7) = \vee\, [((0 \wedge 1), (0.5 \wedge 0.6), (1 \wedge 0)] = 0.5$
for z = 8: $f(8) = \vee\, [(0 \wedge 0.6), (0.5 \wedge 0)] = 0$
for z = 9: $f(9) = 0 \wedge 0 = 0$

Finally, we obtain the representation of P+Q on the figure below:

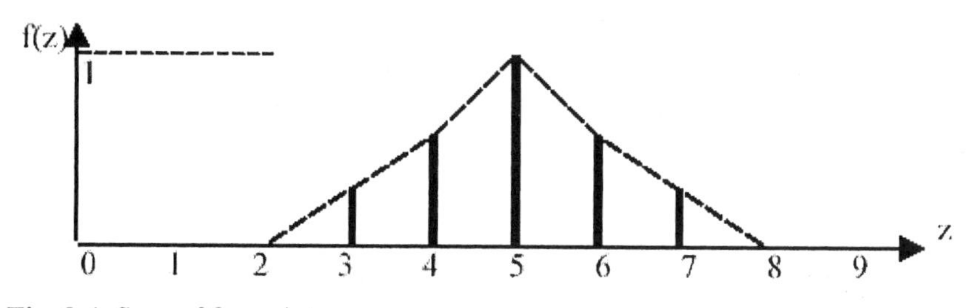

Fig. 2.4. Sum of fuzzy integers

If x_0 and y_0 are the mean values of P and Q respectively, we see that $x_0 + y_0$ is the mean value of the fuzzy number P+Q. Presently, this fuzzy number signifies "z is approximatively equal to 5".
For the multiplication, the z values are given in the following table:

x \ y	1	2	3	4	5
0	0	0	0	0	0
1	1	2	3	4	5
2	2	4	6	8	10
3	3	6	9	12	15
4	4	8	12	16	20

for z = 0: $g(0) = \vee\,[(0 \wedge 0), (0 \wedge 0.6), (0 \wedge 1), (0 \wedge 0.6), (0 \wedge 0)] = 0$
for the other z values: $g(1) = 0$, $g(2) = g(3) = 0.5$, $g(4) = 0.6$, $g(5) = 0$, $g(6) = 1$, $g(8) = 0.6$, $g(9) = 0.5$, $g(10) = 0$, $g(12) = 0.5$, $g(15) = g(16) = g(20) = 0$.

hence the representation of P•Q:

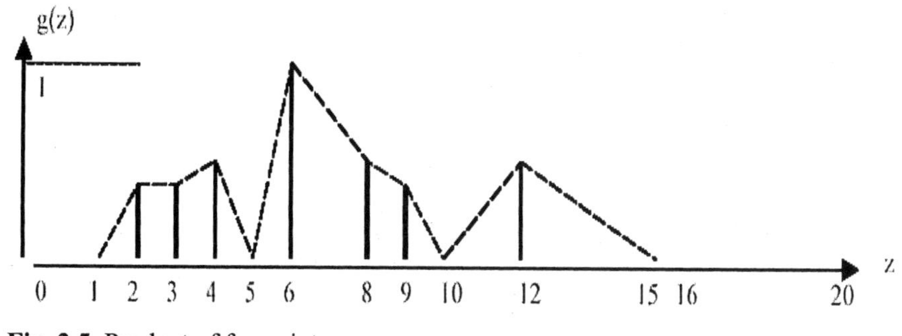

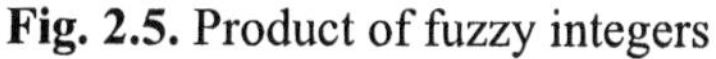

Fig. 2.5. Product of fuzzy integers

We notice that P•Q is normalized but not convex, it is not a fuzzy number, but a fuzzy quantity. The • operator cannot be directly applied to fuzzy numbers when the universe X is a set of discrete values, the resulting fuzzy subsets may no longer be convex and therefore no longer considered as fuzzy numbers.

2.3.2 Other fuzzy logical connectives

Except for the non-contradiction and excluded middle laws, Zadeh's operators preserve the structure of the classical sets theory. Most general fuzzy logical connectives such as the triangular norms and conorms have been defined even if it means getting off the structure.

A *triangular norm*, sometimes called *t-norm*, is a general operator, denoted by T, used for indicating the fuzzy logical "and".

Let $p(x)Tq(x) \in [0,1]$ be the degree of truth of P(x) and Q(x), T must satisfy the following conditions (Bouchon-Meunier 1995, p 39):

1. commutativity: $p(x)Tq(x) = q(x)Tp(x)$
2. associativity: $p(x)T(q(x)Tr(x)) = (p(x)Tq(x))Tr(x)$
   where r(x) is the degree of truth of R(x)
3. isotony: $p(x) \le r(x)$ and $q(x) \le s(x) \Rightarrow p(x)Tq(x) \le r(x)Ts(x)$
   where s(x) is the degree of truth of S(x)
4. neutrality for 1: $p(x)T1 = 1Tp(x) = p(x)$

The most frequently used t-norms are (Fodor and Roubens 1994, pp. 7-8):

$$p(x)T^1q(x) = p(x) \wedge q(x) \tag{2.14}$$

$$p(x)T^2q(x) = p(x) \cdot q(x) \tag{2.15}$$

$$p(x)T^3q(x) = [p(x) + q(x) - 1] \vee 0 \tag{2.16}$$

$$p(x)T^4q(x) = \begin{cases} p(x) \wedge q(x) & \text{if } p(x) + q(x) > 1 \\ 0 & \text{otherwise} \end{cases} \tag{2.17}$$

$$p(x)T^5q(x) = \begin{cases} p(x) \wedge q(x) & \text{if } p(x) \vee q(x) = 1 \\ 0 & \text{otherwise} \end{cases} \tag{2.18}$$

In addition to conditions (1), (2), (3) and (4), any t-norm T verifies properties $0T0 = 0$ and $1T1 = 1$.

Corresponding to the t-norms class, a general class of operators for the fuzzy logical "or" is defined analogously; it is the *triangular* or *t-conorms* class.

A t-conorm, denoted by $\perp$, must satisfy the preceding conditions except for (4). The neutrality condition is defined here in relation to the 0 value, that is to say: $p(x) \perp 0 = 0 \perp p(x) = p(x)$. Moreover, any t-conorm $\perp$ verifies properties $0 \perp 0 = 0$ and $1 \perp 1 = 1$. By considering the duality relation (2.8.b) between the min-max operators, we obtain:

$$1 - ([1 - p(x)] \wedge [1 - q(x)]) = p(x) \vee q(x) \tag{2.19}$$

where $\vee$ is a t-conorm. By substituting the $\wedge$ and $\vee$ operators the most general ones, that is T and $\perp$ respectively, then (2.19) can be used to transform any t-norm T into a t-conorm $\perp$. According to this association, we deduce (Bonissone and Decker 1986) the following t-conorms which correspond to the (2.14)… (2.18) t-norms:

$$p(x) \perp^1 q(x) = p(x) \vee q(x) \tag{2.20}$$

$$p(x) \perp^2 q(x) = p(x) + q(x) - p(x) \,.\, q(x) \tag{2.21}$$

$$p(x) \perp^3 q(x) = [p(x) + q(x)] \wedge 1 \tag{2.22}$$

$$p(x) \perp^4 q(x) = \begin{cases} p(x) \vee q(x) & \mathit{if}\ p(x) + q(x) < 1 \\ 1 & \mathit{otherwise} \end{cases} \tag{2.23}$$

$$p(x) \perp^5 q(x) = \begin{cases} p(x) \vee q(x) & \mathit{if}\ p(x) \wedge q(x) = 0 \\ 1 & \mathit{otherwise} \end{cases} \tag{2.24}$$

Let us note that the complement operator used in (2.19) is a particular negation N such that $N[p(x)] = 1 - p(x)$. Although this negation is very commonly used in practice, there are other ones, for instance (Fodor and Roubens 1994, pp 3-4):

$$N[p(x)] = 1 - [p(x)]^2 \tag{2.25}$$

$$N[p(x)] = \begin{cases} 1 & \mathit{if}\ p(x) = 0 \\ 0 & \mathit{if}\ p(x) > 0 \end{cases} \tag{2.26}$$

$$N[p(x)] = \begin{cases} 1 & \mathit{if}\ p(x) < 1 \\ 0 & \mathit{if}\ p(x) = 1 \end{cases} \tag{2.27}$$

$$N[p(x)] = [1 - p(x)] \,/\, [1 + \lambda\, p(x)] \,, \qquad \lambda > -1 \tag{2.28}$$

More generally, an operator N satisfying the following conditions is a *negation*:

1. $N(0) = 1$ and $N(1) = 0$
2. $p(x) \geq q(x) \Rightarrow N[p(x)] \leq N[q(x)]$

A negation is *strict* if the inequalities in (2) are strict inequalities. Furthermore, if the condition $N\{N[p(x)]\} = p(x)$ is satisfied, then N is said to be *involutive*. We see that the complement operator is a strict and involutive negation. For this category of negation, the duality relation (2.19) extended to the T and $\perp$ operators is written:

$$N\{N[p(x)]\ \mathrm{T}\ N[q(x)]\} = p(x) \perp q(x) \tag{2.29.a}$$

Conversely:

$$N\{N[p(x)] \perp N[q(x)]\} = p(x)\ \mathrm{T}\ q(x) \tag{2.29.b}$$

These general relations show how the t-norms and t-conorms classes are related in a sense of logical duality. Nevertheless, these connectives are not the only way to express the "and" and the "or". A certain number of authors have suggested combining the truth values through the medium of aggregating procedures frequently used in statistics such as arithmetic or geometric means (Zimmermann and Zysno 1980, 1983; Dubois and Grabisch 1994). Here we shall only mention one interesting dual pair of these connectives (called *averaging operators*) due to Werners (1988). The first one, denoted A, concerns the fuzzy "and", the second, denoted $\square$, is the expression of the fuzzy "or". A distinctive feature of these operators is that they combine the minimum and maximum operators, respectively, with the arithmetic mean. Given $\gamma \in [0,1]$, we have:

$$p(x)\ \mathsf{A}\ q(x) = \gamma\ [p(x) \wedge q(x)] + ½\ [\ (1 - \gamma)\ [p(x) + q(x)]\ ] \tag{2.30.a}$$

$$p(x) \square q(x) = \gamma\ [p(x) \vee q(x)] + ½\ [\ (1 - \gamma)\ [p(x) + q(x)]\ ] \tag{2.30.b}$$

If $\gamma = 0$, then $p(x)\ \mathsf{A}\ q(x) = p(x) \square q(x) = ½\ [p(x) + q(x)]$. Inversely, $\gamma = 1$ implies $p(x)\ \mathsf{A}\ q(x) = p(x) \wedge q(x)$ and $p(x) \square q(x) = p(x) \vee q(x)$. It is clear that the parameter γ indicates the degree of nearness of the A and $\square$ operators to the logical meaning of "and" and "or" in the max-min fuzzy logic.

The question arises of how to fix the value of γ within [0,1] ? In other words, do we have to favour the max-min logic (γ near to 1) or have a high regard for the "aggregating" fuzzy logic (γ near to 0) ?

The question can be broached axiomatically (Bellman and Giertz 1973), but the choice of an operator is essentially a matter of context. It mainly depends upon the real-world situation which is to be represented. As far as the applications are concerned, the estimation process of truth values plays an important part in the choice of operators. If the values are estimated with rather unbiased data, it is possible to use averaging operators without any difficulty. But if the degrees of truth are subjective estimates (to assess the beauty of a landscape for instance), we have to regard these estimates as ordinal values and the max-min operators seem to be suitable for the occasion (truth values are only compared, not aggregated in statistical formula).

2.4 Decision-making and evaluation in a fuzzy context

We consider here two simple models within the framework of the max-min fuzzy logic. The first one is due to Bellman and Zadeh (1970), it concerns the decision-making process. The second model worked out by Fustier (1994, 2000) proposes a fuzzy "aggregation" index and applies to the evaluation field.

2.4.1 Optimal fuzzy decision: the Bellman and Zadeh's model

In this well-known model, the universe X represents a set of alternatives denoted x and called *actions*. Corresponding to properties p and q respectively, the fuzzy subsets $P = \{(x / p(x)) \mid x \in X\}$ and $Q = \{(x / q(x)) \mid x \in X\}$ are said to be the *fuzzy objective* and *fuzzy constraint*.

Example: $X = \{a, b, c, d, e\}$ is a set of job applicants in a certain company. This one is searching for "a good economist" (property p) provided that the person in question is "capable of working as a team" (property q). Under these conditions, the fuzzy objective is the fuzzy subset of job applicants who are good economists, for instance $P = \{(a / 0.8), (b / 1), (c / 0.5), (d / 0.4), (e / 0.6)\}$. In the same way, the fuzzy constraint is the fuzzy subset of job applicants who are capable of working as a team, for instance $Q = \{(a / 0.6), (b / 0.6), (c / 0.7), (d / 0.8), (e / 0.1)\}$.

The fuzzy subset D such that $D = P \cap Q$ represents the *decision space*. By definition, D regroups the feasible solutions, that is actions which belong both to the fuzzy objective and the fuzzy constraint. Let d(x) be the membership degree of x to D, we know that $d(x) = p(x) \wedge q(x)$. In the Bellman and Zadeh context, a decision is the act of selecting a specific action which is feasible (element of the decision space): the decision is said to be *optimal* if this action corresponds to the maximum of the objective. In other words, an optimal fuzzy decision consists in selecting the action denoted x_0 which has the highest membership degree in the decision set, that is:

$$d(x_0) = \vee \, [p(x) \wedge q(x) \mid x \in X] \tag{2.31}$$

Remark : x_0 is not always the only solution.

Here we have $D = \{(a / 0.6), (b / 0.6), (c / 0.5), (d / 0.4), (e / 0.1)\}$, thus: $x_0 = a = b$.

The procedure can be extended to any number n of objectives and any number m of constraints, then: $d(x_0) = \vee\, [p_1(x) \wedge \ldots \wedge p_n(x) \wedge q_1(x) \wedge \ldots \wedge q_m(x) \mid x \in X]$.

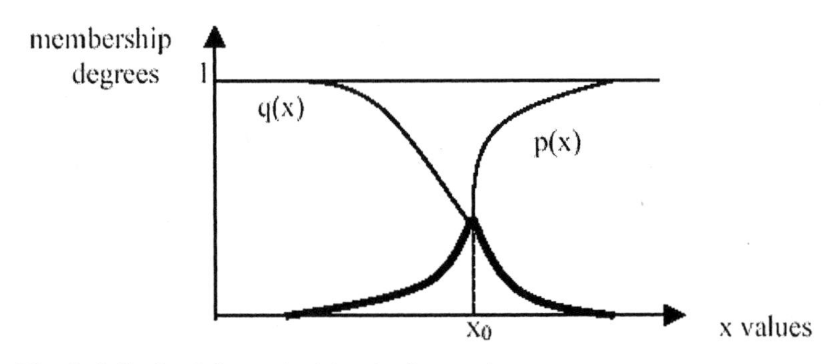

Fig. 2.6. Optimal fuzzy decision in the continuous case

Obviously, the set of actions X can be a set of values. For instance (see Fig. 2.6), the board of directors is trying to find the dividend to be paid to the shareholders. It must be "attractive" for the shareholders (objective, p(x) is increasing). But, the dividend has to be "modest" because of the investment planning of the company (constraint, q(x) is decreasing).

2.4.2 "Fuzzy" aggregation in evaluation problems.

We consider here a set of objects, denoted i, as for example countries that we have to evaluate according to a roughly defined concept like wealth (or its opposite, poverty).

The first step of the evaluating process relies on making the concept of trying to divide the latter into a list of attributes as exhaustive as possible clear. These attributes, denoted j, must be non-redundant and possess different weights denoted $\pi(j)$. If we consider for instance the concept of wealth, we can obtain:

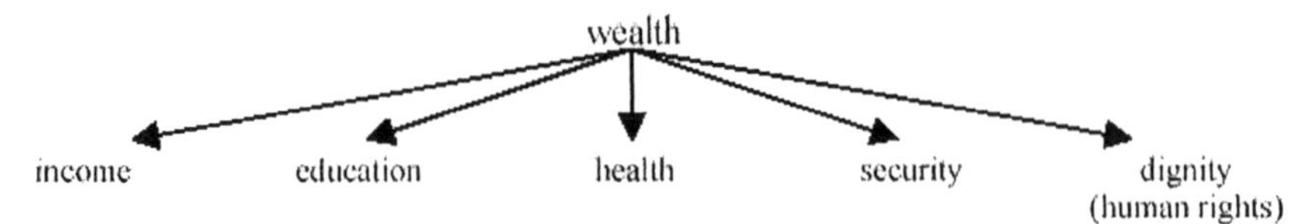

Fig. 2.7. Division of a fuzzy concept into attributes

An attribute is a less vague notion than the initial concept, but it maintains a certain degree of imprecision (from what level of income can we regard a person as well-to-do ? Is as rich as Croesus ? Is the absence of war enough to assert that the people of a country are safe ?). For this reason, the evaluations of the objects on each attribute and the coefficients of importance of these attributes can be considered as truth degrees of fuzzy propositions:

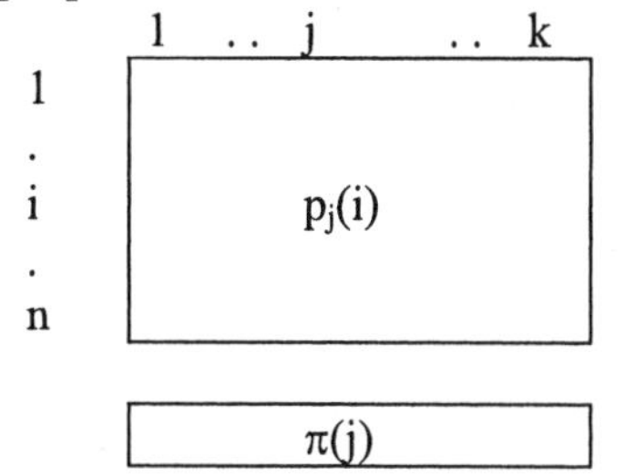

$p_j(i) \in [0,1]$ evaluation of the object i on the attribute j

$\pi(j) \in [0,1]$ coefficient of importance of j.

Fig. 2.8. Data

By definition, $p_j(i)$ is the truth degree of the fuzzy proposition: "i possesses j" and $\pi(j)$ represents the truth degree of the fuzzy proposition "j is important". An attribute with a coefficient of importance equal to 1, is called a *fundamental* attribute; it is assumed that at least one of the attributes is fundamental. Let us note that the vector of coefficients of importance represents the evaluations assigned to an "ideal" object (it possesses j exactly according to the importance of j in the evaluation problem).

If we have to evaluate countries on the first attribute, i.e. according to the monetary wealth (income) and if we can obtain the gross domestic product per capita for each country i, it is possible to consider the formula:

$$p_1(i) = \begin{cases} 0 & \text{if } y(i) < y(-) \\ y(i)/y(+) & \text{otherwise} \end{cases}$$

with y(i) = GDP per capita of i, y(-) = GDP per capita corresponding to the subsistence level, and y(+) = GDP per capita of the richest country in the world.

In case of lack of statistical data concerning purely qualitative attributes (such as "dignity" or the coefficients of importance), we must directly estimate the $p_j(i)$ and $\pi(j)$ in the interval [0,1].

Under these conditions, we wish to define an operator g which assigns a value $g(i) \in [0,1]$ to each object i. Let us observe that g(i) shows how much i fits with the initial concept of evaluation. In the previous example, g(i) is the truth degree of the fuzzy proposition: "the country i is wealthy". By definition, 1 - g(i) is the degree of truth of the proposition: "the country i is not wealthy". Taking into account a concept like wealth or its opposite (poverty) is equally relevant since the fuzzy complementation enables switching from one concept to the other.

There is a wealth of literature on fuzzy aggregation (Dubois and Prade 1985; Mizumoto 1989a, 1989; Fodor and Roubens 1992, 1994; Dubois and Grabisch 1994). Presently, we have to deal with degrees of truth that are more subjective qualitative estimates than objective numerical data (measures). From this statement of fact it follows that the great majority of the compiled operators (the averaging or *compensatory* operators) must be ignored because their theoretical foundations are in no way different from other traditional statistical operators like means. However, we have to admit that operators which are fully compatible with the max-min fuzzy logic are very rare, the well-known operators of this category are the *weighted maximum*:

$$s(i) = \vee [\, p_j(i) \wedge \pi(j) \mid j = 1 \ldots k] \tag{2.32}$$

and the *weighted minimum* operators (Dubois and Prade 1986):

$$s(i)' = \wedge [\, p_j(i) \vee (1 - \pi(j)\,) \mid j = 1 \ldots k] \tag{2.33}$$

Let us note that the weighted minimum does not possess concrete meaning in an evaluation problem (because of the non-importance coefficients $1- \pi(j)$). The weighted maximum formula seems to be appropriate here, but it appears to be too "optimistic": an evaluation equal to 1 given on a fundamental attribute will suffice to obtain a maximum value of the operator, that is 1. We can see this result in table 2.2 where two objects (a and b) and seven attributes (1, 2...) are considered: we obtain $s(a) = 1$ although we have zero evaluations for all the attributes except for $j = 4$.

To find a solution for that, a differential of *discordance* $r_j(i)$ on each j is calculated between the profile of a given object i and the profile of the ideal object (ie the vector of the coefficients of importance):

$$r_j(i) = \begin{cases} 0 & \text{if } p_j(i) \geq \pi(j) \\ \pi(j) - p_j(i) & \text{otherwise} \end{cases} \tag{2.34}$$

We see that $r_j(i) \in [0,1]$ with $r_j(i) = 1$ if j is fundamental and $p_j(i) = 0$. Following the example of the weighted maximum formula, the max-operator is used to summarize the differentials of discordance. Let r(i) be the index of discordance of i, we have:

$$r(i) = \vee [\, r_j(i) \mid j = 1 \ldots k] \tag{2.35}$$

It is clear that $r(i) \in [0,1]$.

An index of *concordance*, denoted t(i), is obtained by the negation of the discordance notion:

$$t(i) = 1 - r(i) \tag{2.36}$$

Table 2.2 Weighted maximum and discordance calculation

j →	1	2	3	4	5	6	7		
$p_j(a)$ →	0	0	0	1	0	0	0		
$p_j(b)$ →	0.9	0.9	0.9	1	0.9	0.9	0.9		
$\pi(j)$ →	1	1	1	1	0.8	0.7	0.7		
$p_j(a) \wedge \pi(j)$ →	0	0	0	1	0	0	0	1	= s(a)
$p_j(b) \wedge \pi(j)$ →	0.9	0.9	0.9	1	0.8	0.7	0.7	1	= s(b)
$r_j(a)$ →	1	1	1	0	0.8	0.7	0.7	1	= r(a)
$r_j(b)$ →	0.1	0.1	0.1	0	0	0	0	0.1	= r(b)

We get t(a) = 0 (a is not in concordance with the ideal object) and t(b) = 0.9 (b is well in concordance with the ideal object). Finally, a fuzzy "aggregation" operator is given by:

$$g(i) = s(i) \wedge t(i) \tag{2.37}$$

According to the max-min fuzzy logic, the $\wedge$-operator is used for connecting the two indexes, but in the applications we can stretch the rules and prefer a more "synthetical" operator such as:

$$g(i) = [s(i) + t(i)] / 2 \tag{2.38}$$

With (2.37) we obtain g(a) = 0 and g(b) = 0.90. With (2.38), we get g(a) = 0.50 and g(b) = 0.95. Remark: from (2.36) and (2.35) we have $t(i) = 1 - \vee [r_j(i) \mid j = 1 \ldots k]$. By using the duality relation (2.8.b), we obtain: $1 - \vee [r_j(i) \mid j = 1 \ldots k] = \wedge [1 - r_j(i) \mid j = 1 \ldots k]$. Finally, we can also calculate the concordance index according to:

$$t(i) = \wedge [1 - r_j(i) \mid j = 1 \ldots k] \tag{2.39}$$

Such a procedure was applied for evaluating the environmental sensibility of tourist zones in the region of Corsica (Fustier and Serra 2001).

From the preceding example, we obtain:

Table 2.3 Using duality relation to calculate the concordance index

j →	1	2	3	4	5	6	7		
$1 - r_j(a)$ →	0	0	0	1	0.2	0.3	0.3	0	= t(a)
$1 - r_j(b)$ →	0.9	0.9	0.9	1	1	1	1	0,9	= t(b)

References

Bellman RE, Giertz M (1973) On the Analytic Formalism of the Theory of Fuzzy Sets. Information Sciences 5:149-156

Bellman RE, Zadeh LA (1970) Decision-making in a fuzzy environment. Management Science 17:141-164

Bonissone PP, Decker KS (1986) Selecting Uncertainty Calculi and Granularity: An Experiment in Trading-Off Precision and Complexity. In: Kanal and Lemmer (eds), Uncertainty in Artificial Intelligence, Amsterdam, pp 217-247

Bouchon-Meunier B (1995) La logique floue et ses applications. Editions Addison – Wesley France, Paris

Dubois D, Grabisch M (1994) Agrégation multicritère et optimisation. In: Logique floue, ARAGO 14 – OFTA. Masson, Paris, pp 179-199

Dubois D, Prade H (1979) Fuzzy Real Algebra: Some Results. Fuzzy Sets and Systems 2:327-348

Dubois D, Prade H (1980) Fuzzy Sets and Systems: Theory and Applications. Academic Press, New York

Dubois D, Prade H (1985) A Review of Fuzzy Sets Aggregation Connectives. Information Sciences 36:85-121

Dubois D, Prade H (1986) Weighted Minimum and Maximum Operations. Fuzzy Sets Theory. Information Sciences 39:12-28

Dubois D, Prade H (1991) On the ranking of ill-known values in possibility theory. Fuzzy Sets and Systems 43:311-317

Fodor J, Roubens M (1992) Aggregation and Scoring Procedures in Multicriteria Decision Making Methods. In: FUZZ - IEEE'92 (Actes de la 1ère Conférence Internationale sur les Systèmes Flous), San Diego, pp 13-21

Fodor J, Roubens M (1994) Fuzzy Preference Modelling and Multicriteria Decision Support. Kluwer Academic Publishers, Dordrecht

Fustier B (1994) Approche qualitative d'un problème d'évaluation. Journal de Mathématiques du Maroc 2:81-86

Fustier B (2000) Evaluation, prise de décision et logique floue. Economie appliquée 1:155-174

Fustier B, Serra D (2001) Sensibilité des sites touristiques. Une approche fondée sur la logique floue. Teoros 3:45-53

Lukasiewicz J (1928) Elements of Mathematical logic. Course given at the University of Warsaw, edited by Pergamon-Polish Scientific Publisher in 1963

Mizumoto M (1989a) Pictoral Representations of Fuzzy Connectives: Cases of t – norms, t –conorms and Averaging Operators. Fuzzy Sets and Systems 31:217-242

Mizumoto M (1989b) Pictoral Representations of Fuzzy Connectives: Cases of Compensatory Operators and Self-dual Operators. Fuzzy Sets and Systems 32:45-79

Ponsard C (1981a) An application of fuzzy subsets theory to the analysis of the consumer's spatial preferences. Fuzzy Sets and Systems 5:235-244

Ponsard C (1981b) L'équilibre spatial du consommateur dans un contexte imprécis. Sistemi Urbani 3:107-133

Ponsard C (1982) Partial spatial equilibria with fuzzy constraints. Journal of Regional Science 22:159-175

Ponsard C (1988) Les espaces économiques flous. In: Ponsard C (ed) Analyse économique spatiale, PUF, Paris, pp 355-389

Ponsard C, Fustier B (1986) Fuzzy Economics and Spatial Analysis. Collection de l'IME – CNRS: 32, Librairie de l'Université (diffuseur), Dijon

Werners B (1988) Aggregation Models in Mathematical Programming. In: Mitra G (ed) Mathematical Models for Decision Support, Berlin, New York, London, Paris, pp 295-319

Zadeh LA (1965) Fuzzy Sets. Information and Control 8:338-353

Zimmermann HJ (1991) Fuzzy set theory and its applications, Second edition. Kluwer Academic Publishers, Norwell

Zimmermann HJ (1993) The German Fuzzy Boom. Les clubs CRIN, La Lettre 4:3-4

Zimmermann HJ, Zysno P (1980) Latent Connectives in Human Decision Making. Fuzzy Sets and Systems 4:37-51

Zimmermann HJ, Zysno P (1983) Decisions and Evaluations by Hierarchical Aggregation of Information. Fuzzy Sets and Systems 10:243-260

3 An Axiomatic Approach to Multidimensional Poverty Measurement via Fuzzy Sets

Satya R Chakravarty[1]

Indian Statistical Institute

3.1 Introduction

Poverty has been in existence for many years and continues to exist in a large number of countries in the World. Therefore, targeting of poverty alleviation remains an important policy issue in many countries. To understand the threat that the problem of poverty poses it is necessary to know the dimension of poverty and the process through which it seems to be deepened. In this context an important question is: how to measure the poverty level of a society and its changes.

In a pioneering contribution, Sen (1976) conceptualized the poverty measurement problem as involving two exercises: (i) the identification of the poor and (ii) aggregation of the characteristics of the poor into an overall indicator that quantifies the extent of poverty. In the literature, the income method has been used mostly to solve the first problem. It requires specification of a poverty line representing the income necessary for a subsistence standard of living. A person is said to be poor if his income falls below the poverty line. On the aggregation issue, Sen (1976) criticized two crude indicators of poverty, the head count ratio (the proportion of persons with incomes below the poverty line) and the income gap ratio (the difference between the poverty line and the average income of the poor, expressed as a proportion of the poverty line), because they remain unaltered under a redistribution of income between two poor persons and the former also does not change if a poor person becomes poorer due to a reduction in his income. Sen (1976) also characterized axiomatically a more sophisticated index of poverty[2].

---

[1] I am grateful to Sabina Alkire and Jacques Silber for bringing some important references to my attention and making them available to me.

[2] Several contributions suggested alternatives and variations of the Sen index. See, for example, Takayama (1979), Blackorby and Donaldson (1980), Kakwani (1980a, 1980b), Clark et al. (1981), Chakravarty (1983a, 1983b, 1983c, 1997), Thon (1983), Foster et al. (1984), Haagenars (1987) and Shorrocks (1995).

However, the well-being of a population, and hence its poverty, which is a manifestation of insufficient well-being, is a multidimensional phenomenon and should therefore depend on both monetary and non-monetary attributes or components. It is certainly true that with a higher income or consumption budget a person may be able to improve the position of some of his non-monetary attributes of well-being. But it may happen that markets for certain non-monetary attributes do not exist. One such example is a public good like flood control or malaria prevention program in an underdeveloped country. Therefore, it has often been argued that income as the sole attribute of well-being is inappropriate and should be supplemented by other attributes, e.g., housing, literacy, life expectancy at birth, nutritional status, provision of public goods etc.

We can provide further justifications for viewing the poverty measurement problem from a multidimensional perspective. In the basic needs approach, advocated by development economists, development is regarded as an improvement in the array of human needs, not just as growth of income alone (Streeten 1981). There is a debate about the importance of low income as a determinant of under nutrition (Lipton and Ravallion 1995) and often it is argued that the population's failure to achieve a desirable nutritional status should be regarded as an indicator of poverty (Osmani 1992). In the capability-functioning approach, where a functioning is what a person "succeeds in doing with the commodities and characteristics at his or her command" (Sen 1985, p.10) and capabilities indicate a person's freedom with respect to functionings (Sen 1985, 1992), poverty is regarded as a problem of functioning failure. Functionings here are closely approximated by attributes like literacy, life expectancy, clothing, attending social activities etc. The living standard is then viewed in terms of the set of available capabilities of the person to function. An example of a multidimensional index of poverty in terms of functioning failure is the human poverty index suggested by the UNDP (1997). It aggregates the country level deprivations in the living standard of a population for three basic dimensions of life, namely, decent living standard, educational attainment rate and life expectancy at birth. Chakravarty and Majumder (2005) axiomatized a generalized version of the human poverty index using failures in an arbitrary number of dimensions of life.

In view of the above, in contrast to the income method, it has often been assumed in the literature that each person is characterized by a vector of basic need attributes (see, for example, Sen 1987, 1992; Ravallion 1996; Bourguignon and Chakravarty 1999, 2003; Atkinson 2003), and a direct method of identification of poor checks if the person has "minimally acceptable levels" (Sen 1992, p. 139) of different basic needs. Therefore, the direct method views poverty from a multidimensional perspective, more precisely, in terms of shortfalls of attribute quantities from respective threshold levels. These threshold levels are determined independently of the attribute distributions. A person is said to be poor with respect to an attribute if his consumption of the attribute falls below its minimally acceptable level. "In an obvious sense the direct method is superior to the income method, since the former is not based on particular assumptions of consumer behavior which may or may not be accurate" (Sen 1981, p. 26). If direct information on different attributes is not available, one can adopt the income method, "so that the income method is at most a second best" (Sen 1981, p. 26).

While the direct and income methods differ substantially in certain respects, they have one feature in common: each individual in the population must be counted as either poor or non-poor. The prospect of an intermediate situation is not considered by them. However, it is often impossible to acquire sufficiently detailed information on income and consumption of different basic needs and hence the poverty status of a person is not always clear cut. For instance, the respondents may be unwilling to provide exact information on income and consumption levels. There can be a wide range of threshold limits for basic needs which co-exist in reasonable harmony. The likelihood that relevant information is missing suggests that there is a degree of ambiguity in the concept of poverty. Now, if there is some ambiguity in a concept, "then a precise representation of that ambiguous concept must *preserve* that ambiguity" (Sen 1997, p. 121). Zadeh (1965) introduced the notion of fuzzy set with a view to tackling problems in which indefiniteness arising from a sort of ambiguity plays a fundamental role. Thus, given that the concept of poverty itself is vague, the poverty status of a person is intrinsically fuzzy. This shows that a fuzzy set approach to poverty measurement is sufficiently justifiable.

Fuzzy set theory –based approaches to the measurement of poverty has gained considerable popularity recently (see, for example, Cerioli and Zani 1990; Blaszczak-Przybycinska 1992; Dagum et al. 1992; Pannuzi and Quaranta 1995; Shorrocks and Subramanian 1994; Cheli and Lemmi 1995; Balestrino 1998; Betti and Verma 1998; Qizilbash 2002)[3].

However, a rigorous discussion on desirable axioms for a multidimensional poverty index in a fuzzy environment has not been carried out in the literature. The purpose of this Chapter is to fill this gap. We also investi-

[3] For applications of fuzzy set to inequality measurement, see Basu (1987) and Ok (1995). Fuzzy set theory is also helpful in analyzing the valuations of functioning vectors and capability sets (see, for example, Balestrino 1994; Balestrino and Chiappero Martinetti 1994; Chiappero Martinetti 1994, 1996, 2004; Casini and Bernetti 1996; Baliamoune 2003; Alkire 2005).

gate how a variety of multidimensional poverty indices suggested recently (see, for example, Chakravarty et al. 1998; Bourguignon and Chakravarty 1999, 2003; Tsui 2002) can be reformulated in a fuzzy structure. These are referred to as fuzzy multidimensional poverty indices.

The Chapter is organized as follows. The next section begins by defining a fuzzy membership function that determines a person's poverty status in a dimension. A characterization of a particular membership function is also presented in this section Sect. 3.3 offers appropriate fuzzy reformulations of the axioms for a multidimensional poverty index. Sect. 3.4 shows how the conventional multidimensional poverty indices can be extended in a fuzzy framework. Finally, Sect. 3.5 concludes.

3.2 Fuzzy Membership Function

We begin by assuming that for a set of *n*-persons, the *ith* person possesses an *k*-vector $(x_{i1}, x_{i2}, ..., x_{ik}) = x_i \in R_+^k$ of attributes, where R_+^k is the non-negative orthant of the *k*-dimensional Euclidean space R^k. The *jth* coordinate of the vector x_i specifies the quantity of attribute *j* possessed by person *i*. The vector x_i is the *ith* row of an $n \times k$ matrix $X \in M^n$, where M^n is the set of all $n \times k$ matrices whose entries are non-negative real numbers. The *jth* of column $x_{.j}$ of $X \in M^n$ gives the distribution of attribute *j* (*j* = *1, 2, ..., k*) among the *n* persons. Let $M = \bigcup_{n \in N} M^n$, where *N* is the set of all positive integers. For any $X \in M$, we write *n (X)* (or *n*) for the associated population size.

In the conventional set up, the poverty status of person *i* for attribute *j* may be represented by a dichotomous function $\mu^*(x_{ij})$, which maps x_{ij} into either zero or one, depending on whether he is non-poor or poor in the attribute, that is, whether $x_{ij} \geq z_j$ or $x_{ij} < z_j$, where $z_j > 0$ is the minimally acceptable or threshold level of attribute *j*. To allow for fuzziness in the poverty status, we consider a more general membership function $\mu_j : R_+^1 \to [0,1]$ for attribute *j*, where $\mu(x_{ij})$ indicates the degree of confidence in the statement that person *i* with consumption level x_{ij} of attribute *j* is possibly poor with respect to the attribute. Thus, μ_j is a generalized characteristic function, that is, one which varies uniformly between zero and one, rather than assuming just two values of zero and one (Zadeh

1965; Chakravarty and Roy 1985). We assume here that μ_j depends on x_{ij} only. One can also consider a more general formulation where μ_j depends on the entire distribution (Cheli and Lemmi 1995). Since μ_j^* declares the poverty status of a person in dimension j unambiguously, we refer to it as a crisp membership function.

Now, let $m_j > 0$ be the quantity of attribute j at or above which a person is regarded as non-poor with certainty with respect to the attribute, that is, if $x_{ij} \geq m_j$, then person i is certainly non-poor in dimension j. (See Cerioli and Zani 1990 and Shorrocks and Subramanian 1994 for a similar assumption in the context of income based fuzzy poverty measurement). For instance, for life expectancy m_j can be taken as the age level 60. Likewise, for the income dimension it can be the level of mean per capita income. We assume here that m_j coincides with one of the x_{ij} values. For example, if a person with the mean level of attribute j, η_j, is considered as certainly non-poor in the attribute, then m_j can be taken as the minimum value of x_{ij} which is at least as large as η_j. That is, m_j $=\min\{x_{ij}\}$, where $i \in \{1,2,....,n\}$ and x_{ij} - η_j ≥ 0. Thus, we can say that the poverty extent of x_{ij}, as measured by μ_j, is zero if $x_{ij} \geq m_j$, that is, $\mu_j(x_{ij}) = 0$ if $x_{ij} \geq m_j$. Similarly if $x_{ij} = 0$ then the poverty level associated with x_{ij} is maximal, and hence $\mu_j(0) = 1$. Furthermore, a reasonable presumption is that a rise in x_{ij} decreases the possibility of person i's being poor in attribute j. Hence μ_j is assumed to be decreasing over $(0, m_j)$. It is also assumed to be continuous. The above properties of μ_j can now be summarized as follows:

$$\begin{aligned} \mu_j(x_{ij}) &= 1 \quad \text{if} \quad x_{ij} = 0, \\ \mu_j(x_{ij}) &= 0 \quad \text{if} \quad x_{ij} \geq m_j . \end{aligned} \tag{3.1}$$

It is decreasing over the interval $(0, m_j)$ and continuous everywhere. We write μ for the vector $(\mu_1, \mu_2, \ldots, \mu_k)$. Let A be the set of vectors of membership functions of the form μ.

An example of a suitable fuzzy membership function for attribute *j* is:

$$\mu_j(x_{ij}) = \begin{cases} 1 & \text{if } x_{ij} = 0 \\ \left(\dfrac{m_j - x_{ij}}{m_j}\right)^{\theta_j} & \text{if } x_{ij} \in (0, m_j) \\ 0 & \text{if } x_{ij} \geq m_j \end{cases} \tag{3.2}$$

where $\theta_j \geq 1$ is a parameter.

It satisfies all the conditions laid down (3.1). It is an individualistic function in the sense that it depends only on x_{ij} and treats m_j as a parameter.

Given μ_j, let $S_{\mu_j}(X)$ (or, simply S_{μ_j}) be the set of persons who are possibly poor in dimension *j* in $X \in M^n$, where $n \in N$ is arbitrary, that is:

$$S_{\mu_j}(X) = \{i \in \{1, 2, \ldots, n\} \mid \mu_j(x_{ij}) > 0\} \tag{3.3}$$

Attribute *j* will be called possibly meager or certainly non-meager for person *i* according as $i \in S_{\mu_j}(X)$ or $i \notin S_{\mu_j}(X)$. Person *i* is referred to as certainly non-poor if $x_{ij} \geq m_j$ for all $j = 1, 2, \ldots k$, that is, if $i \notin S_{\mu_j}(X)$ for all *j*.

It will now be worthwhile to characterize a fuzzy membership function. Such a characterization exercise will enable us to understand the membership function in a more elaborate way through the axioms used in the exercise. The following axioms are proposed for a general membership function $\mu_j : R^1_+ \to [0,1]$ for attribute *j*.

$(A1)$ Homogeneity of Degree Zero: μ_j is homogeneous of degree zero.

$(A2)$ Linear Decreasingness: For any $x_{ij} \in [0, m_j)$ and $c_{ij} \in [0, m_j - x_{ij})$, $\mu_j(x_{ij}) - \mu_j(x_{ij} + c_{ij}) = \dfrac{c_{ij}}{m_j}$.

$(A3)$ Continuity: μ_j is continuous on its domain.

$(A4)$ Maximality: $\mu_j(0) = 1$.

$(A5)$ Independence of Non-meager Attribute Quantities: For all $x_{ij} \geq m_j$, $\mu_j(x_{ij}) = k$, where k is a constant.

$(A1)$ ensures that μ_j remains unaltered under equi-proportionate variations in quantities of attribute j. $(A2)$ makes a specific assumption about the decreasing of the membership function. It says that the extent of reduction in the membership function resulting from an increase in x_{ij} by c_{ij} is the fraction c_{ij}/m_j. It is weaker than the decreasing assumption of the membership function over $[0, m_j)$. A membership function may as well decrease non-linearly. For instance, if $\theta_j > 1$, μ_j in (3.2) decreases at an increasing rate. $(A3)$ means that μ_j should vary in a continuous manner with respect to variations in attribute quantities. $(A4)$ specifies that μ_j should achieve its maximal value 1 when the level of the attribute is zero. Finally, $(A5)$ shows insensitivity of μ_j to the attribute quantities of the persons who are certainly non-poor in the attribute through the assumption that the value of the membership function on $[m_j, \infty)$ is a constant. Thus, instead of assuming that the membership function takes on the value zero on $[m_j, \infty)$, we derive it as an implication of more primitive axioms.

Proposition 1: The only membership function that satisfies axioms $(A1)-(A5)$ is:

$$\mu_j(x_{ij}) = \begin{cases} 1 & \text{if } x_{ij} = 0 \\ \left(\dfrac{m_j - x_{ij}}{m_j}\right) & \text{if } x_{ij} \in (0, m_j) \\ 0 & \text{if } x_{ij} \geq m_j \end{cases} \tag{3.4}$$

Proof: In view of $(A1)$, we have $\mu_j(x_{ij}) = \mu_j(\frac{x_{ij}}{m_j})$. Hence $(A2)$ becomes:

$$\mu_j(\frac{x_{ij}}{m_j}) - \mu_j(\frac{x_{ij} + c_{ij}}{m_j}) = \frac{c_{ij}}{m_j}$$

Since in the above equation, $x_{ij} \in [0, m_j)$ is arbitrary, we can interchange the roles of x_{ij} and c_{ij} in it and derive that:

$$\mu_j(\frac{c_{ij}}{m_j}) - \mu_j(\frac{c_{ij}+x_{ij}}{m_j}) = \frac{x_{ij}}{m_j}$$

These two equations imply that:

$$\mu_j(\frac{x_{ij}}{m_j}) - \mu_j(\frac{c_{ij}}{m_j}) = \frac{c_{ij}}{m_j} - \frac{x_{ij}}{m_j}.$$

Letting c_{ij} =0 in the above expression , we get:

$$\mu_j(\frac{x_{ij}}{m_j}) = \mu_j(0) - \frac{x_{ij}}{m_j},$$

from which in view of $(A4)$ it follows that:

$$\mu_j(\frac{x_{ij}}{m_j}) = \frac{m_j - x_{ij}}{m_j}.$$

Applying $(A1)$ to the above form of μ_j and using $(A3)$, we note that $\mu_j(m_j) = 0$. This along with $(A5)$ reveals that $k = 0$. Hence $\mu_j(x_{ij}) = 0$ for all $x_{ij} \geq m_j$. This establishes the necessity part of the proposition. The sufficiency is easy to check. Δ

Proposition 1 thus characterizes axiomatically the linear sub-case of the membership function in (3.2).

3.3 Properties for a Fuzzy Multidimensional Poverty Index

In this section we lay down the postulates for a fuzzy multidimensional poverty index $P: M \times A \to R^1$. For all $n \in N$, the restriction of P on $M^n \times A$ is denoted by P^n. For any $X \in M^n, \mu \in A, P^n(X;\mu)$ gives the extent of possible poverty (poverty, for short) level associated with X.

Sen (1976) suggested two basic postulates for an income poverty index. These are: (i) the monotonicity axiom, which requires poverty to increase under a reduction in the income of a poor person, and (ii) the transfer axiom, which demands that poverty should increase if there is a transfer of income from a poor person to anyone who is richer. Following Sen (1976) several other axioms have been suggested in the literature. (See, for example, Sen 1979; Foster 1984; Foster et al. 1984; Donaldson and Weymark 1986; Seidl 1988; Chakravarty 1990; Foster and Shorrocks 1991; Zheng 1997). Multidimensional generalizations of different postulates proposed for an income poverty index have been introduced, among others, by

Chakravarty et al. (1998), Bourguignon and Chakravarty (1999, 2003) and Tsui (2002). The axioms we suggest below for an arbitrary P are fuzzy variants of the axioms presented in Chakravarty et al. (1998), Bourguignon and Chakravarty (1999, 2003) and Tsui (2002).

Focus (FOC): For all $n \in N; X, \hat{X} \in M^n; \mu \in A;$ if $S_{\mu_j}(X) = S_{\mu_j}(\hat{X})$, $1 \leq j \leq k$ and $x_{ij} = \hat{x}_{ij}$ for all $i \in S_{\mu_j}(X), 1 \leq j \leq k$, then:

$$P^n(X;\mu) = P^n(\hat{X};\mu).$$

Normalization (NOM): For all $n \in N; X \in M^n; \mu \in A; j \in \{1,2,\ldots,k\}$, if $S_{\mu_j}(X) = \phi$, the empty set, then $P^n(X;\mu) = 0$.

Monotonicity (MON): For all $n \in N; X, \hat{X} \in M^n; \mu \in A$; if $x_{rl} = \hat{x}_{rl}$ for all $r \in \{1,2,\ldots,n\} - \{i\}, \ell \in \{1,\ldots,k\}$; $x_{il} = \hat{x}_{il}$ for all $\ell \in \{1,\ldots,k\} - \{j\}$ and $x_{ij} > \hat{x}_{ij}$ where $i \in S_{\mu_j}(\hat{X})$, then $P^n(X;\mu) < P^n(\hat{X};\mu)$.

Transfers Principle (TRP): For all $n \in N; X, \hat{X} \in M^n; \mu \in A$, if X is obtained from $\hat{X}$ by multiplying $\hat{X}_p$ by a bistochastic matrix B and $B\hat{X}_p$ is not a permutation of the rows of $\hat{X}_p$, then $P^n(X;\mu) < P^n(\hat{X};\mu)$, where $\hat{X}_p$ is the matrix of attribute quantities of possibly the poor in $\hat{X}$, given that the bundles of attributes of the rich remain unaffected[4].

Principle of Population (POP): For all $n \in N; X \in M^n; \mu \in A$, $P^n(X;\mu) = P^{hn}(\hat{X};\mu)$ where $\hat{X}$ is the h – fold replication of X.

Symmetry (SYM): For all $n \in N; X \in M^n; \mu \in A$:

$$P^n(X;\mu) = P^n(\Pi X;\mu), \text{ where } \Pi \text{ is an } n \times n \text{ permutation matrix.}$$

Subgroup Decomposability (SUD): For $X^1, X^2, \ldots, X^h \in M$ and $\mu \in A$:

[4] An $n \times n$ matrix is called a bistochastic matrix if its entries are non-negative and each of its rows and columns sums to one. A bistochastic matrix is called a permutation matrix if there is exactly one positive entry in each row and column.

$P^n(X^1, X^2, \ldots, X^h; \mu) = \sum_{i=1}^{h} \frac{n_i}{n} P^{n_i}(X^i; \mu)$, where n_i is the population size corresponding X^i and $n = \sum_{i=1}^{h} n_i$.

Continuity (CON): For all $n \in N;\ \mu \in A;\ P^n(X;\mu)$ is continuous on M^n.

Increasingness in Membership Functions (IMF): For all $n \in N; X \in M^n$, $\mu, \mu' \in A$ if $\mu_h = \mu'_h$ for all $h \in \{1,\ldots,k\} - \{j\}, S_{\mu_j}(X) = S_{\mu'_j}(X)$ and $\mu_j(x_{ij}) > \mu'_j(x_{ij})$ for all $i \in S_{\mu_j}$, then $P^n(X;\mu') < P^n(X;\mu)$.

Non-poverty Growth (NPG): For all $n \in N; X \in M^n; \mu \in A$ if $\hat{X}$ is obtained from X by adding a certainly non-poor person to the society, then $P^{n+1}(\hat{X};\mu) < P^n(X;\mu)$.

Scale Invariance (SCI): For all $n \in N; X \in M^n; \mu \in A$:
$P^n(X\Omega;\mu) = P^n(X;\mu)$, where Ω is the diagonal matrix:
diag $(\omega_1, \omega_2, \ldots, \omega_k)$, $\omega_j > 0$ for all $j = 1, \ldots, k$.

FOC, which has a similar spirit to $(A5)$, states that, given the population size, the poverty index depends only on the attribute quantities of the persons who are possibly poor in different dimensions. Thus, if a person is certainly non-poor with respect to an attribute, then giving him more of this attribute does not change the intensity of poverty, even if he is possibly poor in the other attributes. Clearly, **FOC** rules out trade off between the two attributes of a person who is possibly poor with respect to one but certainly non-poor with respect to the other. Thus, if life expectancy and composite good are the two attributes, more life expectancy in the domain in which it is certainly non-meager is of no use if the composite good is possibly meager. This, however, does not exclude the possibility of a trade off if both the attributes are possibly meager for a person. **NOM** is a cardinality property of the poverty index. It says that if all persons in a society are certainly non-poor, then the index value is zero. According to **MON**, poverty decreases if the condition of a poor improves. **MON** includes the possibility that the beneficiary may become certainly non-poor in the dimension concerned.

To understand **TRP**, let us recall a result from the literature on inequality measurement. Of two income distributions u and v of a given total over a given population size n, where u is not a permutation of v ,the former can be obtained from the latter through a sequence of rank preserving progressive transfers transferring incomes from the better off persons to those who are worse off if and only if $u = vB$ for some bistochastic matrix B of order n (Kolm 1969; Dasgupta et al. 1973). In the multidimensional context, Kolm (1977) showed that the distribution of a set of attributes summarized by some matrix X is more equal than another matrix $\hat{X}$ (whose rows are not identical) if and only if $X = E\hat{X}$, where E is some bistochastic matrix and X cannot be derived from $\hat{X}$ by permutation of the rows of $\hat{X}$. Intuitively, multiplication of $\hat{X}$ by a bistochastic matrix makes the resulting distribution less concentrated. Following Kolm (1977), the analogous property applied to the set of possibly poor persons is **TRP**. It simply says that there is less possible poverty under X than under $\hat{X}$ if the former is obtained from the latter by redistributing the attributes of the possibly poor using some bistochastic transformation.

Under **POP**, if an attribute matrix is replicated several times, then poverty remains unchanged. Since by replication we can transform two different sized matrices into the same size, **POP** is helpful for inter-temporal and interregional poverty comparisons. **SYM** demands anonymity. Any characteristic other than the quantities in different dimensions under consideration, for instance, the names of the individuals, is immaterial to the measurement of poverty. **CON,** which is similar to $(A3)$, ensures that minor changes in attribute quantities will not give rise to an abrupt jump in the value of the poverty index. Therefore, a continuous poverty index will not be oversensitive to minor observational errors on basic need quantities.

SUD says that if a population is divided into several subgroups, say h, defined along ethnic, geographical or other lines, then the overall poverty is the population share weighted average of subgroup poverty levels. The contribution of subgroup i to overall poverty is $n_i P^{n_i}(X^i;\mu)/n$ and overall poverty will precisely fall by this amount if poverty in subgroup i is eliminated.

$(n_i P^{n_i}(X^i;\mu)/\ nP^n(X;\mu))100$ is the percentage contribution of subgroup i to total poverty. Each of these statistics is useful to policy-makers because they become helpful for isolating subgroups of the population that are more susceptible to poverty (see Anand 1997; Chakravarty 1983a; Foster et al. 1984; Foster and Shorrocks 1991).

Between two identical communities, the one with higher membership function of an attribute should have a higher poverty because of higher possibility of individuals' being poor in that dimension. This is what **IMF** demands. A poverty index will be called μ**-monotonic** if it satisfies **IMF**. According to **NPG** poverty should decrease if a person who is certainly rich joins the society. Thus, under **FOC**, **NPG** says that the poverty index is a decreasing function of the population size (see Kundu and Smith 1983; Subramanian 2002; Chakravarty et al. 2005). Finally, **SCI,** which parallels $A1$, means that the poverty index is invariant under scale transformations of attribute quantities, that is, it is homogeneous of degree zero. Hence it should be independent of the units of measurement of attributes. Thus, if life expectancy is measured in months instead of in years, the level of poverty remains unchanged.

We will now consider a property which takes care of the essence of multidimensional measurement through correlation between attributes. By taking into account the association of attributes, as captured by the degree of correlation between them, this property also underlines the difference between single and multidimensional poverty measurements. To illustrate the property, consider the two-person two-attribute case, where both the attributes are possibly meager for these persons. Suppose that $x_{11} > x_{21}$ and $x_{12} < x_{22}$. Now, consider a switch of attribute 2 between the two persons. This switch increases the correlation between the attributes because person 1 who had more of attribute 1 has now more of attribute 2 too and that is why we refer to it as a correlation increasing switch between two possibly poor persons. Formally, we have:

Definition 1: For any $n \geq 2; X \in M^n; \mu \in A; j,h \in \{1,2,\ldots..,k\}$, suppose that for some $i,t \in S_{\mu_j}(X) \cap S_{\mu_h}(X)$, $x_{ij} < x_{tj}$ and $x_{th} < x_{ih}$. $\hat{X}$ is then said to be obtained from X by a correlation increasing switch between two possibly poor persons if $(i)\hat{x}_{ij} = x_{tj}, (ii)\hat{x}_{tj} = x_{ij}, (iii)\hat{x}_{rj} = x_{rj}$ for all $r \neq i,t$ and $(iv)\hat{x}_{rs} = x_{rs}$ for all $s \neq j$ and for all r.

If the two attributes are substitutes, that is, if one attribute compensates for the lack of another for a person who is possibly poor in both dimensions, then the switch should increase poverty. This is because the richer of the possibly poor is getting even better in the attributes which correspond to the similar aspect of poverty after the rearrangement. After the switch the poorer person is less able to compensate the lower quantity of one attribute by the quantity of the other. Indeed, the switch just defined does not modify the marginal distribution of each attribute but it reduces the extent to which the lack of one attribute may be compensated by the availability of the other. An analogous argument will establish that poverty should decrease under a correlation increasing switch if the two attributes are complements. (For more detailed arguments along this line, see Atkinson and

Bourguignon 1980; Bourguignon and Chakravarty 2003). We state this principle formally for substitutes as:

Increasing Poverty Under Correlation Increasing Switch (IPC): For all $n \in N; \mu \in A; X \in M^n$, if $\hat{X}$ is obtained from X by a correlation increasing switch between two possibly poor persons, then $P^n(X;\mu) < P^n(\hat{X};\mu)$ if the two attributes are substitutes.

The corresponding property which demands poverty to decrease under such a switch when the attributes are complements is denoted by **DPC**. If a poverty index does not change under a correlation increasing switch, then it treats the attributes as "independents".

3.4 The Subgroup Decomposable Fuzzy Multidimensional Poverty

3.4.1 Poverty Indices

The objective of this section is to discuss the subgroup decomposable family of fuzzy multidimensional poverty indices. The necessity for a subgroup decomposable index arose from practical considerations. The use of such an index allows policy-makers to design effective, consistent national and regional anti-poverty policies.

Repeated application of **SUD** shows that we can write a subgroup decomposable index as:

$$P^n(X;\mu) = \frac{1}{n}\sum_{i=1}^{n} p(x_i;\mu) \tag{3.4}$$

where $n \in N; X \in M^n$ and $\mu \in A$ are arbitrary. Since $p(x_i;\mu)$ depends only on person *i's* consumption of the attributes, we call it "individual poverty function". If we define $p(x_i;\mu)$ as the weighted average of grades of membership of individual i across dimensions, that is, if $p(x_i;\mu) = \sum_{j=1}^{k} \delta_j \mu_j(x_{ij})$, where $0 < \delta_j < 1$ and $\sum_{j=1}^{k} \delta_j = 1$, then P^n in (3.4) becomes:

$$P^n(X;\mu) = \frac{1}{n}\sum_{j=1}^{k} \delta_j \sum_{i \in S_{\mu_j}} \mu_j(x_{ij}) \tag{3.5}$$

The weight δ_j may be assumed to reflect the importance that we attach in our aggregation to dimension j. It may also be assumed as reflecting the importance that the government assigns for alleviating poverty for that dimension. Since $\sum_{i \in S_{\mu j}} \mu_j(x_{ij})$ gives the cardinality of the fuzzy set of the poor in the *jth* attribute (Dubois and Prade 1980, p. 30), P^n in (3.5) is a weighted average of the proportions of possibly poor persons across dimensions. If μ_i coincides with the crisp membership function μ_j^*, then the index in (3.5) becomes a weighted average of the proportions of persons who are poor in different dimensions.

Alternatively we may interpret the formula as follows. $\mu_j(x_{ij})$ can be regarded as the extent of deprivation felt by person i for being included in the set of persons who are possibly poor in attribute j. As his quantity of consumption of the attribute increases, deprivation decreases and $\mu_j(m_j) = 0$ shows the absence of this feeling at the level m_j. Therefore, P^n is the population average of the weighted average of dimension –wise individual deprivations.

Defining $\frac{1}{n} \sum_{i \in S_{\mu j}} \mu_j(x_{ij})$ as the possible poverty level associated with attribute j and denoting it by $P^n(x_{.j}; \mu_j)$, we can rewrite P^n in (3.5) in a more compact way as:

$$P^n(X, \mu) = \sum_{j=1}^{k} \delta_j P_j^n(x_{.j}; \mu_j) \tag{3.6}$$

This shows that $P^n(X; \mu)$ can also be viewed as a weighted average of attribute-wise (possible) poverty values. We refer to this property as "Factor Decomposability". The percentage contribution of dimension j to total fuzzy poverty is $\left(\delta_j P^n(x_{.j}; \mu_j) / P^n(X; \mu)\right)100$. The elimination of poverty for the *jth* dimension will lower community poverty by the amount $\delta_j P^n(x_{.j}; \mu_j)$.

We can use the two decomposability postulates to construct a two-way poverty profile and to calculate each attribute's poverty within each subgroup. This type of micro breakdown will help us to identify simultaneously the population subgroup(s) as well as attribute(s) for which poverty levels are severe and formulate appropriate antipoverty policies.

It will now be worthwhile to examine the behavior of P^n given by (3.5) with respect to the axioms stated in Sect. 3.3. These axioms conveniently translate into constraints on the form of μ_j. Evidently, P^n in (3.5) is focused, normalized, monotonic, symmetric, population replication invariant, μ-monotonic, continuous and correctly responsive to non-poverty growth. It satisfies **SCI** if and only if for each *j*, μ_j is homogeneous of degree zero, a condition fulfilled by the form given in (3.2). It is transfer preferring, that is, **TRP** holds if and only if μ_j is strictly convex over $(0, m_j)$, $1 \le j \le k$, (see Marshall and Olkin 1979, p. 433). This means that the decline in the possibility of poverty with increase in quantities of attributes is greatest at the lowest levels of the attribute. The membership function defined in (3.2) satisfies the convexity condition if $\theta_j \ge 2$. Finally, because of additivity across attributes it remains unchanged under a correlation increasing switch. We summarize these observations on the behavior of P^n as follows:

Proposition 2: The subgroup decomposable fuzzy multidimensional poverty index given by (3.5) satisfies the **Focus**, **Normalization**, **Monotonicity**, **Principle of Population, Symmetry**, **Continuity**, **Increasingness in Membership Functions** and **Non-Poverty Growth** axioms. It fulfills the **Scale Invariance** axiom if and only if the membership functions for different attributes are homogeneous of degree zero. It meets the **Transfers Principle** axiom if and only if for each j, μ_j is strictly convex on the relevant part of the domain. Finally, it remains unchanged under a correlation increasing switch between two possibly poor persons.

To illustrate the general formula in (3.5), suppose that the membership function is of the form (3.2). In this case the index is:

$$P_\theta^n(X;\mu) = \frac{1}{n}\sum_{j=1}^{k}\delta_j \sum_{i \in S_{\mu j}} \left(1 - \frac{x_{ij}}{m_j}\right)^{\theta_j} \tag{3.7}$$

where $\theta = (\theta_1, \theta_2, \ldots, \theta_k)$, which reflect different perceptions of poverty. This is a fuzzy counterpart to the multidimensional generalization of the Foster – Greer – Thorbecke (FGT) (1984) index considered by Chakravarty et al. (1998) and Bourguignon and Chakravarty (2003). For a given X, P_θ^n increases as θ_j increases, $1 \le j \le k$. For $\theta_j = 1$, for all *j*, P_θ^n becomes:

$$P_{\theta}^{n}(X;\mu)=\frac{1}{n}\sum_{j=1}^{k}\delta_{j}H_{j}I_{j} \tag{3.8}$$

where I_j is the average of the grades of membership of the persons in S_{μ_j} when $\theta_j=1$, that is, $I_j = \sum_{i\in S_{\mu j}}(m_j - x_{ij})/q_j m_j$, with q_j being the cardinality of S_{μ_j} and $H_j = q_j/n$ is the fuzzy head-count ratio in dimension *j*. Thus, for a given H_j, an increase in I_j, say, due to a reduction of x_{ij}, increases the index.

If $\theta_j = 2$ for all *j*, P_{θ}^{n} can be written as:

$$P_{\theta}^{n}(X;\mu)=\sum_{j=1}^{k}\delta_{j}H_{j}\left[I_{j}^{2}+(1-I_{j})^{2}C_{j}^{2}\right] \tag{3.9}$$

where $C_j^2 = \sum_{i\in S_{\mu j}}(x_{ij}-\rho_j)^2/q_j\rho_j^2$ is the squared coefficient of variation of the distribution of attribute *j* among those for whom it is possibly meager, with $\rho_j = \sum_{i\in S_{\mu j}} x_{ij}/q_j$ being the mean of the distribution. Now, C_j^2 is an index of inequality of the concerned distribution. Clearly, given I_j and H_j, P_{θ}^{n} in (3.9) reduces as C_j reduces, say through a transfer from a less possibly poor to a more possibly poor. Thus, the decomposition in (3.9) shows that the poverty index is related in a positive monotonic way with the inequality levels of the possibly poor in different dimensions.

An alternative of interest arises from the following specification of the membership function:

$$\mu_j(x_{ij}) = 1-\left(\frac{x_{ij}}{m_j}\right)^{c_j} \tag{3.10}$$

where for all j, $1\le j\le k, c_j\in(0,1)$. It satisfies all the conditions laid down in (3.1) along with homogeneity of degree zero and strict convexity. The associated poverty index is:

$$P_{c}^{n}\ (X;\mu)=\frac{1}{n}\sum_{j=1}^{k}\delta_{j}\sum_{i\in S_{\mu j}}\left[1-\left(\frac{x_{ij}}{m_j}\right)^{c_j}\right] \tag{3.11}$$

where $c=(c_1,c_2,\ldots,c_m)$. This index is a fuzzy version of the multidimensional extension of the subgroup decomposable single dimensional Chakravarty (1983b) index suggested by Chakravarty et al. (1998). Given X, P_c^n is increasing in c_j for all j. For $c_j=1$, the index coincides with the particular case of P_θ^n when $\theta_j=1,\ 1\le j\le k$. On the other hand as $c_j\to 0$ for all j, $P_c^n\to 0$. As c_j decreases over the interval (0, 1), P_c^n becomes more sensitive to transfers lower down the scale of distribution along dimension j.

We may also consider a logarithmic formulation of the membership function that fulfils all conditions stated in (3.1):

$$\mu_j(x_{ij})=\frac{\log\left(1+e^{\lambda_j(m_j-x_{ij})/m_j}\right)-\log 2}{\log\left(1+e^{\lambda_j}\right)-\log 2} \tag{3.12}$$

where $\lambda_j>0$ is a parameter. The corresponding additive poverty index turns out to be:

$$P_\lambda^n(X;\mu)=\frac{1}{n}\sum_{j=1}^{k}\delta_j\sum_{i\in S_{\mu j}}\frac{\log\left(1+e^{\lambda_j(m_j-x_{ij})/m_j}\right)-\log 2}{\log\left(1+e^{\lambda_j}\right)-\log 2} \tag{3.13}$$

where λ is the parameter vector $(\lambda_1,\lambda_2,\ldots,\lambda_k)$. P_λ^n can be regarded as a fuzzy sister of the multidimensional generalization of the Watts (1967) poverty index characterized by Tsui (2002). The parameter λ_j determines the curvature of the poverty contour. An increase in λ_j for any j makes the fuzzy poverty contour more convex to the origin. If $\lambda_j\to 0$ for all j, then $P_\lambda^n\to 0$. In the particular case when $\theta_j=\lambda_j=1$ for all j, the ranking of two attribute matrices $X,\hat{X}\in M^n$ by P_θ^n will be the same as that generated by P_λ^n. Since P_λ^n is transfer preferring for all $\lambda_j>0$, it satisfies TRP even in this case. But P_θ^n does not fulfill TRP here.

There can be simple non-additive formulations of fuzzy multidimensional extensions of single dimensional subgroup decomposable indices.

They satisfy **SUD** but not factor decomposability. Assuming that θ_j in (3.2) is constant across attributes, say equal to β, one such index can be:

$$P^n_{\alpha\beta}(X;\mu) = \frac{1}{n}\left[\sum_{j=1}^{k}\sum_{i=1}^{n} a_j \mu_j(x_{ij})\right]^{\frac{\alpha}{\beta}} = \frac{1}{n}\left[\sum_{j=1}^{k}\sum_{i\in S_{\mu_j}} a_j\left(\frac{m_j - x_{ij}}{m_j}\right)^{\beta}\right]^{\frac{\alpha}{\beta}} \quad (3.14)$$

where $a_j > 0$ for all j and α is a positive parameter. $P^n_{\alpha,\beta}$ is the fuzzy counterpart to the multidimensional version of the FGT index suggested by Bourguignon and Chakravarty (2003). The interpretation of this index is quite straightforward. The membership functions in various dimensions are first aggregated into a composite membership using a particular value of β and the coefficients a_j. Multidimensional fuzzy poverty is then defined as the average of that composite membership value, raised to the power α, over the whole population. $P^n_{\alpha,\beta}$ satisfies **IPC** or **DPC** depending on whether α is greater or less than β. For $\alpha = 1$, it becomes the weighted sum of order β of the membership grades and for a given $X \in M^n$, it is increasing in β.

We may suggest an alternative to (3.14) using the membership function in (3.11). This form is defined by:

$$T^n_c(X;\mu) = \frac{1}{n}\sum_{i=1}^{n}\left(1 - \prod_{j=1}^{k}(1-\mu_j)\right) = \frac{1}{n}\sum_{i=1}^{n}\left(1 - \prod_{j=1}^{k}\left(\frac{\hat{x}_{ij}}{m_j}\right)^{c_j}\right) \quad (3.15)$$

where $\hat{x}_{ij} = \min(x_{ij}, m_j)$. This is a fuzzy translation of the multidimensional generalization of the Chakravarty (1983b) index developed by Tsui (2002). In (3.15) for each person complements from unity of the grades of membership along various dimensions are subjected to a product transformation which is then averaged over persons after subtracting from its maximum value, that is, 1. Since T^n_c is unambiguously decreasing under a correlation increasing switch between two possibly poor persons, it treats the concerned attributes unambiguously as complements, that is, it satisfies **DPC**.

Given a membership fraction μ_j, there will be a corresponding multidimensional fuzzy poverty index that meets all the postulates considered in Sect. 3.2. These indices will differ only in the manner in which we use μ_j to aggregate membership grades of different persons along different dimensions in an overall indicator.

3.5 Conclusions

This Chapter has explored the problem of replacing the traditional crisp view of poverty status with a fuzzy structure which allows membership of poverty set or the possibility of poverty in different dimensions of life to take any value in the interval [0, 1]. An attempt has been made to establish how standard multidimensional poverty indices might be translated into the fuzzy framework. Suggestions have been made for suitable fuzzy analogues of axioms for a multidimensional poverty index, such as, **Focus**, **Monotonicity**, **Transfers Principle** and **Continuity**. We have also added a condition which requires poverty to increase if the possibility of poverty shifts upward along any dimension.

We will now make a comparison of our index with some existing indices. Assuming that the individual well-being depends only on income, Cerioli and Zani (1990) suggested the use of the arithmetic average of grades of membership of different individuals as a fuzzy poverty index. It "represents the proportions of individuals "belonging" in a fuzzy sense to the poor subset" (Cerioli and Zani 1990, p. 282). Clearly, this index is similar in nature to P'' given by (3.5).

In a multidimensional framework, Cerioli and Zani (1990) introduced a transition zone $x_j^L < x_{ij} \leq x_{ij}^H$ for attribute j over which the membership function declines from 1 to 0 linearly:

$$\mu_j(x_{ij}) = \begin{cases} 1 & \text{if } x_{ij} \leq x_j^{(L)} \\ \dfrac{x_j^{(H)} - x_{ij}}{x_j^{(H)} - x_j^{(L)}} & \text{if } x_{ij} \in \left(x_j^{(L)}, x_j^{(H)}\right] \\ 0 & \text{if } x_{ij} > x_j^{(H)} \end{cases} \tag{3.16}$$

They then defined the membership function for person i as $\dfrac{\sum_{j=1}^{k} \mu_j(x_{ij}) w_j}{\sum_{i=1}^{k} w_i}$, where $w_1, w_2, \ldots, w_k$ represent a system of weights.

In what has been called the "Totally Fuzzy and Relative" approach, Cheli and Lemmi (1995) defined the membership function for attribute j as the distribution function $F(x_{ij})$, normalized (linearly transformed) so as to equal 1 for the poorest and 0 for the richest person in the population:

$$\mu_j\left(x_{ij}\right)=\begin{cases} 1 & if \quad x_{ij}=x_j^{(s)} \\ \mu_j\left(x_j^{(l-1)}\right)+\dfrac{F(x_j^{(l)})-F(x_j^{(l-1)})}{1-F(x_j^{(l)})} & if \quad x_{ij}=x_j^{(l)} \\ 0 & if \quad x_{ij}=x_j^{(1)} \end{cases} \tag{3.17}$$

where $x_j^{(1)}, x_j^{(2)}, \ldots, x_j^{(s)}$ are modalities of dimension j in increasing order with respect to the risk of poverty connected to them.

An alternative specification of the membership function for person i arises if we replace μ_j in (16) by μ_j in (3.17). In either case, as Cerioli and Zani (1990) and Cheli and Lemmi (1995) suggested, under appropriate specification of weights, we can take:

$$C'' = \frac{\sum_{i=1}^{n} \sum_{j=1}^{k} \mu_j(x_{ij})w_j}{n\sum_{i=1}^{k} w_i} \tag{3.18}$$

as an indicator of poverty. Cerioli and Zani (1990) chose $w_j = \log(1/p_j)$, where p_j is the proportion of persons with *jth* poverty symptoms, and Cheli and Lemmi (1995) preferred to use $w_j = \log\left(n/\sum_{i=1}^{n}\mu_j(x_{ij})\right)$. C'' indicates the cardinality of the fuzzy subset of the poor as a proportion of the population size.

An important difference between P'' in (3.5) and C'' is that while P'' is subgroup decomposable, C'' is not. This is because C'' depends on different kinds of rank orders. Precisely, because of this a poverty index based on a Gini type inequality index or welfare function is not subgroup decomposable. Examples are the Sen (1976), Kakwani (1980b) and Thon (1983) indices.

A rank preserving transfer of some quantity of an attribute from a possibly poor to a worse off person will not change the rank orders of the modalities in the concerned dimension. Therefore, satisfaction of the **Transfers Principle** by the general index C'' will depend on the assumption about the weight system. Likewise, a rank preserving reduction in the quantity of an attribute will not change the rank orders of the modalities. Hence a similar argument holds concerning fulfillment of **Monotonicity.** However, C'' is normalized, symmetric, scale invariant (under appropriate choices of modalities) and responds correctly to non-poverty growth. It is continuous for the membership function in (3.16). Continuity for the membership function in (3.17) will hold if F is continuous. To check whether it is population replication invariant, concrete specification of the weight sequence is necessary.

References

Alkire S (2005) Measuring the freedom aspects of capabilities. Global Equity Initiative, Harvard University

Anand S (1997) Aspects of poverty in Malaysia. Review of Income and Wealth 23:1 – 16

Atkinson AB (2003) Multidimensional deprivation: contrasting social welfare and counting approaches. Journal of Economic Inequality 1:51 – 65

Atkinson AB, Bourguignon F (1982) The comparison of multidimensioned distributions of economic status. Review of Economic Studies 49:183 –201

Balestrino A (1994) Poverty and functionings: issues in measurement and public action. Giornale degli economisti e Annali di economia 53:389-406

Balestrino A (1998) Counting the poor in a fuzzy way: the head-count ratio and the monotonicity transfer axioms. Notizie di Politeia 14:77–86

Balestrino A, Chiappero Martinetti E (1994) Poverty, differentiated needs and information. Mimeo, University of Pisa

Baliamoune M (2003) On the measurement of human well-being: fuzzy set theory and Sen's capability approach. Presented at the WIDER conference on Inequality, Poverty and Human Well-being, 30-31 May

Basu K (1987) Axioms for fuzzy measures of inequality. Mathematical Social Sciences 14:69-76

Betti G, Verma VK (1998) Measuring the degree of poverty in a dynamic and comparative context: a multidimensional approach using fuzzy set theory. Working Paper n. 22, Department of Quantitative Methods, University of Siena

Blackorby C, Donaldson D (1980) Ethical indices for the measurement of poverty. Econometrica 58:1053–1060

Blaszczak-Przybycinska I (1992) Multidimensional statistical analysis of poverty in Poland. In: Polish Statistical Association and Central Statistical Office (ed.), Poverty measurement for economies in transition. Warsaw, pp 307-327

Bourguignon F, Chakravarty SR (1999) A family of multidimensional poverty measures. In: Slottje DJ (ed), Advances in econometrics, income distribution and scientific methodology: essays in honor of C. Dagum. Physica–Verlag, Heidelberg, pp 331–344

Bourguignon F, Chakravarty SR (2003) The measurement of multidimensional poverty. Journal of Economic Inequality 1:25–49

Casini L, Bernetti I (1996) Public project evaluation, environment, sustainability and Sen's theory. In: Balestrino A, Carter I (eds) Functionings and capabilities: normative and policy issues. Notizie di Politeia, 12:55-78

Cerioli A, Zani S (1990) A fuzzy approach to the measurement of poverty. In: Dagum C, Zenga M (eds) Income and wealth distribution, inequality and poverty. Springer–Verlag, New York, pp 272-284

Chakravarty SR (1983a) Ethically flexible measures of poverty. Canadian Journal of Economics 16:74–85

Chakravarty SR (1983b) A new index of poverty. Mathematical Social Sciences 6:307–313

Chakravarty SR (1983c) Measures of poverty based on representative income gaps. Sankhya 45:69-74

Chakravarty SR (1990) Ethical social index numbers .Springer–Verlag, New York

Chakravarty SR (1997) On Shorrocks' reinvestigation of the Sen poverty index . Econometrica 65:1241-1242

Chakravarty SR, Kanbur R, Mukherjee D (2005) Population growth and poverty measurement. Social Choice and Welfare: forthcoming

Chakravarty SR, Majumder A (2005) Measuring human poverty: a generalized index and an application using basic dimensions of life and some anthropometric indicators. Journal of Human Development 6:275-299

Chakravarty SR, Mukherjee D, Ranade R (1998) On the family of subgroup and factor decomposable measures of multidimensional poverty. Research on Economic Inequality 8:175–194

Chakravarty SR, Roy R (1995) Measurement of fuzziness: a general approach. Theory and Decision 19:163-169

Cheli B, Lemmi A (1995) A "Totally" fuzzy and relative approach to the multidimensional analysis of poverty. Economic Notes 24:115-134

Chiappero Martinetti E (1994) A new approach to evaluation of well-being and poverty by fuzzy set theory. Giornale degli economisti e annali di economia 53:367-388

Chiappero Martinetti E (1996) Standard of living evaluation based on Sen's approach: some methodological suggestions. Politeia 12:37-53

Chiappero Martinetti E (2005) Complexity and vagueness in the capability approach: strengths or weaknesses? In: Comim F, Qizilbash M, Alkire S (eds) The capability approach in human development: concepts, applications and measurement. Cambridge University Press, Cambridge: forthcoming

Clark S, Hemming R, Ulph D (1981) On indices for the measurement of poverty Economic Journal 91:515-526

Dagum C, Gambassi R, Lemmi A (1992) New approaches to the measurement of poverty. In: Polish Statistical Association and Central Statistical Office (ed) Poverty measurement for economies in transition. Warsaw, pp 201-226

Dasgupta P, Sen AK, Starett D (1973) Notes on the measurement of inequality. Journal of Economic Theory 6:180–187

Donaldson D, Weymark JA (1986) Properties of fixed-population poverty indices. International Economic Review 27:667–688

Dubois D, Prade H (1980) Fuzzy sets and systems. Academic Press, London

Foster JE (1984) On economic poverty measures: a survey of aggregate measures. In: Basmann RL, Rhodes GF (eds) Advances in Econometrics, Vol.3. JAI Press, Connecticut

Foster JE, Greer J, Thorbecke E (1984) A class of decomposable poverty measures. Econometrica 42:761–766

Foster JE, Shorrocks AF (1991) Subgroup consistent poverty indices. Econometrica 59:687–709

Haagenars AJM (1987) A class of poverty measures. International Economic Review 28:593–607

Kakwani NC (1980a) Income inequality and poverty: methods of estimation and policy applications. Oxford University Press, Oxford

Kakwani NC (1980b) On a class of poverty measures. Econometrica 48:437–446

Kolm SC (1969) The optimal production of social justice. In: Margolis J, Guitton H (eds) Public economics. Macmillan, London, pp 145–200

Kolm SC (1977) Multidimensional egalitarianism. Quarterly Journal of Economics 91:1–13

Kundu A, Smith TE (1983) An impossibility theorem on poverty indices. International Economic Review 24:423–434

Lipton M, Ravallion M (1995) Poverty and policy. In: Behrman J, Srinivasan TN (eds) Handbook of development economics, Vol. 1. North Holland, Amsterdam

Marshall AW, Olkin I (1979) Inequalities: theory of majorization and its applications. Academic Press, London

Ok E (1995) Fuzzy measurement of income inequality: a class of fuzzy inequality measures. Social Choice and Welfare 12:115-136

Osmani S (1992) Introduction. In: Osmani SR (ed) Nutrition and poverty .Oxford University Press, Oxford

Pannuzi N, Quaranta AG (1995) Measuring poverty: a study case in an Italian industrial city. In: Dagum C, Lemmi A (eds) Income Distribution, Social Welfare, Inequality and Poverty, Volume 6 of Slottje DJ (ed) Research on Economic Inequality, JAI Press Inc., London, pp 323-336

Qizilbash M (2002) A note on the measurement of poverty and vulnerability in the South African context. Journal of International Development 14:757-772

Ravallion M (1996) Issues in measuring and modeling poverty. Economic Journal 106:1328–1343

Seidl C (1988) Poverty measurement: a survey. In: Bos D, Rose M, Seidl C (eds) Welfare efficiency in public economics. Springer-Verlag, New York

Sen AK (1976) Poverty: on ordinal approach to measurement. Econometrica 44:219–231

Sen AK (1979) Issues in the measurement of poverty. Scandinavian Journal of Economics 81:285-307

Sen AK (1981) Poverty and famines. Clarendon Press, Oxford

Sen AK (1985) Commodities and capabilities. North Holland, Amsterdam

Sen AK (1987) Standard of living. Cambridge University Press, Cambridge

Sen AK (1992) Inequality re-examined. Harvard University Press, Cambridge, MA

Sen AK (1997) On economic inequality, with a substantial annexe by Foster JE. Oxford University Press, Oxford

Shorrocks AF (1995) Revisiting the Sen poverty index. Econometrica 63:1225–1230

Shorrocks AF, Subramanian S (1994) Fuzzy poverty indices. University of Essex

Streeten P (1981) First things first: meeting basic human needs in developing countries. Oxford University Press, Oxford

Subramanian S (2002) Counting the poor: an elementary difficulty in the measurement of poverty. Economics and Philosophy 18:277–285

Takayama N (1979) Poverty income inequality and their measures: Professor Sen's axiomatic approach reconsidered. Econometrica 47:749–759

Thon D (1983) A poverty measure. Indian Economic Journal 30:55-70

Tsui KY (2002) Multidimensional poverty indices. Social Choice and Welfare 19:69–93

UNDP (1997) Human development report. Oxford University Press, Oxford

Watts H (1968) An economic definition of poverty. In: Moynihan DP (ed) On understanding poverty. Basic Books, New York

Zadeh L (1965) Fuzzy sets. Information and Control 8:338-353

Zheng B (1997) Aggregate poverty measures. Journal of Economic Surveys 11:123-162

4 On the Convergence of Various Unidimensional Approaches

Ehud Menirav[1]

The Eitan Berglas School of Economics, Tel-Aviv University

4.1 Introduction

The public debate on poverty often attempts to identify the characteristics of the poor population, explain changes in poverty trends and shape programs to relieve their plight. In the course of these discussions, it is commonly assumed that the poor population and poverty trends are well defined and unambiguous. However, as it turns out, neither premise is well founded and the outcomes of each depend on the specific way poverty is defined and measured.

Several studies have demonstrated the sensitivity of poverty measurement depends on just how we measure it. Hagenaars and De Vos (1988), Glewwe and Van Der Gaag (1990), Abul Naga (1994) and Mercader-Prats (1998) have shown that a population defined as poor on the basis of income is far from being the same population defined on the basis of consumption. Buhmann et al. (1988), Coulter et al. (1992), Atkinson (1992), Slesnick (1993), Banks and Johnson (1994), Jenkins and Cowell (1994) and Burkhauser et al. (1996) have stressed that the choice of an equivalence scale affects the values of poverty indices. Smeeding et al. (1993) have discussed the effect of non-cash income on poverty rates. Finally, Sen (1979), Atkinson (1991) and Ravallion (1994) have dwelled on the theoretical importance of the weighting procedure.

The present study examines poverty in Israel while taking into account the sensitivity of poverty indices and identification of the poor population to the choice of the economic well-being variable (reflecting choice of one poverty dimension) in addition to the equivalence scale and weighting procedure employed. The study, based on 1997 census data, demonstrates

---

[1] This Chapter is based in part on my thesis written at Bar-Ilan University. Some parts of this Chapter were drawn from the author's previous publication in Hebrew (Menirav 2002). The Chapter was written when the author was visiting Bonn University. I wish to thank Jacques Silber for helpful discussions and comments.

these issues by means of a systematic unidimensional approach using a single dataset. Some of the sensitivity results have never been empirically examined in the literature. More particularly, the study considers the empirical effect of the weighting procedure on both poverty indices and the identification of the poor population[2]. To the best of my knowledge, this is the first study to show that local taxes enhance the scope of poverty. Other results reproduce findings that have already been noted in the literature (e.g., the effect of the equivalence scale on poverty indices). Yet, despite the cases cited, the present study covers a wider spectrum of indices and introduces results from a new dataset.

The sensitivity analysis is based on a comparison of 48 distributions derived from Israel's 1997 *Household Expenditures Survey,* containing data on households in urban localities. This is a rich dataset that covers consumption patterns, nutrition level, income level and its composition, housing conditions, and so forth. The 48 distributions, each considered as a separate case, were obtained by combining the distributions of four economic well-being variables, six equivalence scales and two weighting procedures. Hence, each case represents a different combination of these components.

The Chapter is organized as follows. Sect. 4.2 reviews the basic components of the unidimensional approach to poverty measurement applied. Sect. 4.3 analyzes the impact of the definition chosen on the scope of poverty while Sect. 4.4 examines how those definitions affect identification of the poor. Sect. 4.5 concludes the study.

4.2 Basic components of the unidimensional approach

The unidimensional approach to poverty measurement applied in the research is composed of five main elements: types of economic well-being variables, equivalence scale, weighting procedures, poverty line and poverty indices. In this section these elements are briefly reviewed within the context of the present study.

One of the most important issues in poverty measurement is the selection of the economic well-being variable, a factor that captures one specific poverty dimension. When making this decision, we try to ascertain what single variable best measures household welfare. Although the literature has shown some preference for using consumption when computing poverty level (see e.g., Glewwe and Van Der Gaag 1990; Atkinson 1991;

[2] Nonetheless, Danziger and Taussig (1979) and Cowell (1984), examined the empirical effect of the weighting procedure on the extent of inequality.

and McGregor and Nachane 1995), in practice, some definition of income rather than consumption is officially employed in most countries. In this study four economic well-being variables that are considered relevant for poverty measurement in Israel are used: net cash income, net income, net cash income after deduction of municipal taxes and consumption[3].

1. *Net cash income* is defined as the sum of all the household's current sources of income after deduction of income tax, social security and compulsory health insurance premiums.
2. *Net income* is defined as the sum of net cash income in addition to all non-cash income, which is an imputed estimate of income in kind from durable goods (dwellings and vehicles) as well as the value of goods supplied free of charge by the individual's employer or by some other institution. Imputed income from durable goods is the imputed interest and depreciation on vehicles and "alternative rent" (i.e., the sum a household would receive if its residence were rented).
3. *Net cash income after deduction of local (municipal) taxes* is an income concept that does not appear directly in the 1997 Survey of Household Expenditures but was constructed especially for this study.
4. *Consumption* is defined as all household payments for the purchase of products or services, as well as imputed expenditure on the consumption of housing services and vehicles (the purchase of which is defined as investment, not consumption).

Let $N = \{1, 2, \ldots, n\}$ be the set of households represented in an economy and let y_i denote the economic well-being variable of household $i \in N$.

Next consider the issue of equivalence scales, an instrument meant to improve the problem of treating households with different needs[4]. The single (but certainly not only) most important factor that equivalence scales attempt to capture is household size. In this study six equivalence scales are used. Five of those scales correspond to the five different values of the equivalence elasticity (α = 0, 0.25, 0.5, 0.75, and 1) appearing in the equivalence scale proposed by Buhmann et al. (1988) (hereinafter *BES*(α)): $x_i = y_i/(m_i)^{\alpha}$ for $0 \leq \alpha \leq 1$, where m_i is the number of members in household i and x_i is the *adjusted* value of the economic well-being variable enjoyed by the members of that household. Without loss of gener-

[3] Atkinson's (1991) distinction between a concept of poverty based on the net income and one based on net cash income is worth mentioning. The former stresses the standard of living whereas the latter stresses what he calls "the minimum right to resources".

[4] For a general discussion of equivalence scales see Cowell and Mercader-Prats (1999) and Deaton and Muellbauer (1980).

ality, let's assume that households are arranged in ascending order of their adjusted economic well-being variable: $x_1 \leq x_2 \leq \ldots \leq x_n$.

When $\alpha = 0$, the welfare enjoyed by household members is the total value of the economic well-being variable for the household whereas when $\alpha = 1$, it is the per capita value of that variable. Another way of illustrating the difference between the extreme cases, where α equals either 0 or 1, is to say that in the first case we assume that all goods consumed by the household members are "public" goods. This implies that household size has no impact on the level of welfare its members enjoy. In the second case, all goods are "private" goods. In this case, the larger the household, the lower the welfare enjoyed by each member (for a given value of the economic well-being variable). Values of α falling between 0 and 1 correspond to intermediate cases, where the larger the α, the smaller the economies of scale.

The sixth equivalence scale employed in this study is the one actually used in Israel by the National Insurance Institute (hereinafter NII-ES). In order to rank households by their economic situation, Israel's National Insurance Institute divides household income by the number of "standard persons" present. This number is constructed in the following way (the first number is the actual number of persons in the household; the number in parenthesis is the number of standard persons): 1 – (1.25), 2 – (2.00), 3 – (2.65), 4 – (3.20), 5 – (3.75), 6 – (4.25), 7 – (4.75), 8 – (5.20). Every additional "real" person has a marginal weight of 0.4 standard persons.

Regardless of how the equivalence scales are determined, the issue of weighting procedure arises. As shown by Danziger and Taussig (1979), Sen (1979) and Cowell (1984), three alternative weighting approaches may be considered: (i) applying an equal weight to each household (irrespective of its size), (ii) applying the same weight to each person (irrespective of household size) and (iii) applying a weight equal to the number of "equivalent persons" in each household. Sen (1979) claims that the third weighting procedure is not recommended because it results from a conceptual confusion. Between the first two procedures, Sen (1979) prefers the approach giving equal weight to every individual – the approach followed by the majority of researchers – although no consensus has been reached (see for example Ebert (1999), who demonstrates the merits of the third weighting procedure). The distinction between the weighting procedure and the equivalence scale may best be summarized by the questions asked with respect to each. Regarding this issue, we ask: "Which welfare measure best represents the economic well-being of households for the purpose of measuring poverty?" For the weighting procedure we ask: "How should we aggregate the poor population?" In the present study the analysis is limited to two weighting procedures: assigning an equal weight to each household (hereinafter *EWH*) or assigning an equal weight to each individual (hereinafter *EWI*). Let w_i denote the weight of household $i \in N$. Accordingly, $w_i = 1$ for all i when employing *EWH* and $w_i = m_i$ when employing *EWI*.

According to the relative approach, the poverty line is defined as being equal to half the median of the adjusted economic well-being variable. Although, the sensitivity of the poverty line definition is not considered per se (i.e., whether to employ 40% or 60% or any other percentage of the median of the adjusted economic well-being variable), it is worth mentioning that the value of the poverty line is distribution-sensitive because the median changes with the choice of that variable and the weighting procedure.

We can now turn to a brief review of the five poverty indices employed in this study: the head-count ratio, the income gap ratio, the poverty gap ratio, the Foster et al. (1984) index and Sen's (1976) axiomatic index[5]. The most common measure of overall poverty is the head-count ratio (*H*), which is defined as the proportion of poor economic units to overall population: $H = \sum_{i=1}^{q} w_i \big/ \sum_{i=1}^{n} w_i$ where q is the number of poor households.

Another popular measure is the income gap ratio (*I*), which is the distance between the poverty line and the average value of the economic well-being variable for the poor $I = \sum_{i=1}^{q} (z - x_i) \cdot w_i \big/ (z \cdot \sum_{j=1}^{q} w_j) = 1 - \mu_p / z$, where z denotes the poverty line and μ_p the average economic well-being variable for the poor. The poverty gap ratio (*HI*) is defined as the product of the head-count ratio and the income gap ratio, that is:

$$HI = \sum_{i=1}^{q} (z - x_i) \cdot w_i \Big/ \left(\sum_{j=1}^{n} w_j \cdot z\right) = I \cdot H .$$

Most importantly, the previous measures are insensitive to the distribution of income among the poor. The Foster, Greer and Thorbecke (1984) index (hereinafter *FGT*), in contrast, is sensitive to the distribution of income among the poor (where δ is different from both 0 and 1):

$$FGT = \left(1 \Big/ \sum_{i=1}^{n} w_i\right) \cdot \sum_{i=1}^{q} w_i \cdot \left(1 - x_i / z\right)^{\delta} \text{ with } \delta \geq 0.$$

One can observe that when $\delta = 0$, $FGT = H$, and when $\delta = 1$, $FGT = HI$. Naturally, the more general case is that in which δ is different from both 0 and 1; it is in those cases that we turn to the *FGT*. In the current study, the parameter δ is set to equal 0.5.

[5] See Foster (1984), Atkinson (1987), Hagenaars (1987) and Zheng (1997) for extensive surveys of poverty indices and their different properties.

Another distribution-sensitive index is Sen's (1976) axiomatic index (S), expressed as: $S = H \cdot [I + (1 - I) \cdot G]$, where G is the Gini coefficient of the adjusted economic well-being variable calculated for the poor.

4.3 The choice of definition and the scope of poverty

Let us now turn to consider whether the choice of the economic well-being variable, the equivalence scale and the weighting procedure has an impact on poverty index values.

4.3.1 Impact of the weighting procedures

The choice of weighting procedure has a significant impact on the scope of poverty. To see that, let d denote the ratio of a poverty index measured assuming *EWH* (in numerator of d) to the same poverty index measured using *EWI* (in denominator of d), given the same economic well-being variable and equivalence scale for both. That is, cases with $d > 1$ correspond to cases where poverty is higher when assigning an equal weight for each household than when an equal weight is assigned to each individual.

Table 4.1 summarizes the distribution of the ratio d. A total of 120 ratios were computed using all possible combinations of the four economic well-being variables, six equivalence scales and five poverty indices. As the table readily shows, there are only four cases where a poverty index is higher when individuals rather than households are assigned an equal weight (yet, even in these cases, the values of the poverty indices are almost identical). Significantly, in most cases, poverty is higher when using *EWH* than when using *EWI*. Similar conclusions, as far as inequality indices are concerned, were derived by Danziger and Taussig (1979) and are likely to be the consequence of the linkage between household size and its income (despite the assumed importance of the issue, it is not discussed in the present study).

Table 4.1. Frequency distribution of the ratio d

Bin	Frequency
0.96 – 1	4
1 – 1.25	70
More than 1.25	46

4.3.2 Impact of the economic well-being variables

Choice of the variable measuring economic well-being is crucial for its impact on the scope of poverty. The poverty level was highest when employing net cash income after deduction of local taxes. The second-highest poverty level was obtained when analyzing the distribution of net cash income. Poverty was somewhat lower when the net income concept was adopted and lowest when consumption was the basis for measurement. In only 3 out of the 60 groups of 4 economic well-being variables a slightly different ranking was obtained. Similar results – at least as far as the comparison of consumption and current income is concerned – were obtained by Slesnick (1993).

To illustrate and extend these results, consider the Jenkins and Lambert (1997) "three 'I's of poverty" curve (or *TIP* for short). Let $g_i = \sum_{j=1}^{i} \max\{z - x_j, 0\} \cdot w_j \big/ (z \cdot \sum_{l=1}^{n} w_l)$ denote cumulative sum of the normalized poverty gap per capita, and let $p_i = \sum_{j=1}^{i} w_j \big/ \sum_{l=1}^{n} w_l$ denote cumulative population share. A *TIP* curve links g_i to p_i (see Fig. 4.1 for a demonstration based on the data analyzed in this study). The *TIP* curve monotonously increases with respect to p_i and becomes horizontal when $p_i = H$; in addition, at this stage, $g_i = HI$. Moreover, the slope of the ray from the origin of the axes to the critical point where the curve becomes horizontal is equal to the income gap ratio (see the *TIP* consumption curve in Fig. 4.1). The area under the curve equals half of the Shorrocks (1995) modified-Sen index and approaches half of the Thon (1979) index as n tends to infinity.

Jenkins and Lambert (1997, 1998a, 1998b) and Davidson and Duclos (2000) have shown that if one *TIP* curve consistently lies above another, the higher (or "dominating") curve will always correspond to a greater scope of poverty for a family of poverty indices. This family includes all the indices that are replication invariant, increasing Schur-convex functions of the normalized poverty gaps. Included, for example, are the Watts (1968) index, the Clark et al. (1981) index, the Chakravarty (1983) index, the *FGT* index with $\delta \geq 1$, the *HI* index as well as the Shorrocks (1995) modified-Sen index.

Fig. 4.1 shows that the ranking of the curves corresponds to the conclusions derived above. The highest *TIP* curve is that of net cash income after deduction of local taxes; the lowest *TIP* curve is that derived on the basis of consumption. The same ranking of *TIP* curves also holds for small values of p_i although this phenomenon is difficult to discern on the figure.

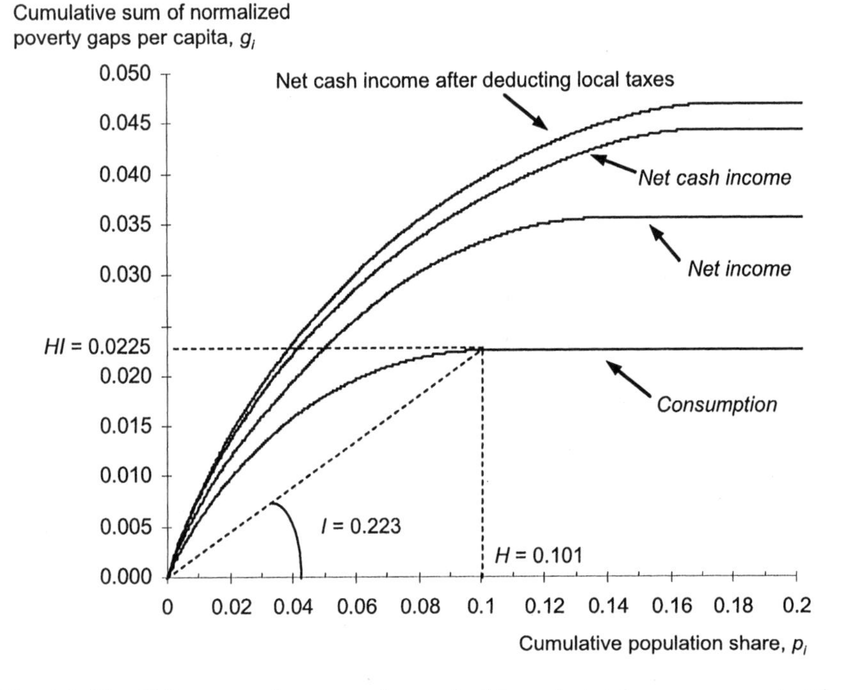

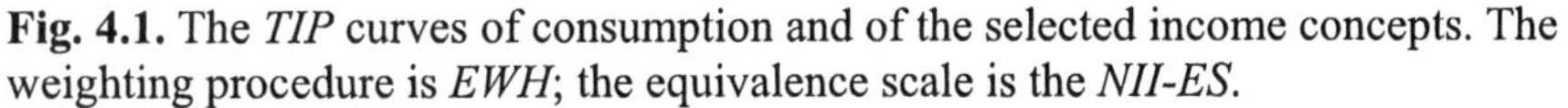

Fig. 4.1. The *TIP* curves of consumption and of the selected income concepts. The weighting procedure is *EWH*; the equivalence scale is the *NII-ES*.

The most interesting and perhaps surprising finding is the effect of local taxes. This result seems to contradict the fact that the local tax rate in Israel is supposed to be lower for households that are more likely to have limited means a priori. In Israel, reductions in local taxes are given, for example, to several categories of the elderly, to individuals who suffer from disabilities (e.g., blind people), to new immigrants, to single-parent families or to families considered needy by the social welfare system. One may therefore wonder why taking local taxes into account induces an increase in poverty.

One can imagine several explanations. Firstly, some households may not take advantage of their right to these reductions, perhaps due to unawareness of its existence. Second, it may be that households whose income is well above the poverty line also benefit from local tax reductions. If such is the case, it can be shown that the scope of poverty tends to increase as a result. To take a simplified example, assume that y_m is median income, T is a lump-sum local tax, the poverty line is equal to half of median income and households whose income is beneath the poverty line or slightly above it benefit from a 60% reduction in local taxes (these assumptions do not alter household rank in the region proximate to the poverty line). The gap between the poverty line and those poor originally having income y decreases from $(0.5y_m - y)$ to $[0.5(y_m - T) - (y - 0.4T)] = (0.5y_m - y) - 0.1T$.[6] However, if households earning an income corresponding to the median level also benefit from a reduction in local taxes, say a 30% reduction, the previously defined gap will *increase* to $[0.5(y_m - 0.7T) - (y - 0.4T)] = (0.5y_m - y) + 0.05T$, resulting in an increase in the scope of poverty.

Third, it is also possible that the reduction in local taxes to poor households is inadequate. Going back to the previous example, assume that households whose income is lower (or slightly higher) than the poverty line benefit from a 40% reduction: The gap then increases from $(0.5y_m - y)$ to $[0.5(y_m - T) - (y - 0.6T)] = (0.5y_m - y) + 0.1T$.[7]

Next, the scope of poverty is compared when employing net cash income and net income. The inclusion of non-cash income in the definition of net cash income (generating what is defined as net income) leads to a reduction in the scope of poverty. This result implies that more considerable differences in cash income than in non-cash income separate households. The data show, for example, that the ratio of average net income per household between the fifth and the second decile (the space where the poverty line is usually located) is equal to 2.06 while the corresponding ratio for non-cash income is only equal to 1.41.

4.3.3 Impact of equivalence scales

The analysis shows that the choice of equivalence scale has a significant impact on the scope of poverty. Poverty indices first decline but later rise with an increase in the value of the equivalence elasticity (α) of $BES(\alpha)$. This U-shaped relationship between poverty level and equivalence scale is clearly illustrated in Fig. 4.2. Similar results were found by Coulter et al. (1992) regarding the poverty indices H, HI and FGT with $\delta = 2$.

---

[6] In such a case most poverty indices will show a reduction in the scope of poverty, sometimes because some households will climb above the poverty line and sometimes because this gap has narrowed. Another way of illustrating this result is to think of such a case as resembling that where there is no local tax, and the population that would otherwise be eligible for a reduction in local taxes will receive a positive transfer in the amount of $0.1T$.

[7] This example can be extended to cases where there is a monotonic non-decreasing local tax. Call T the highest local tax and assume there is a new scheme of local tax reductions. Again, an insufficiently large reduction in local taxes awarded to the poor population may increase the poverty indices.

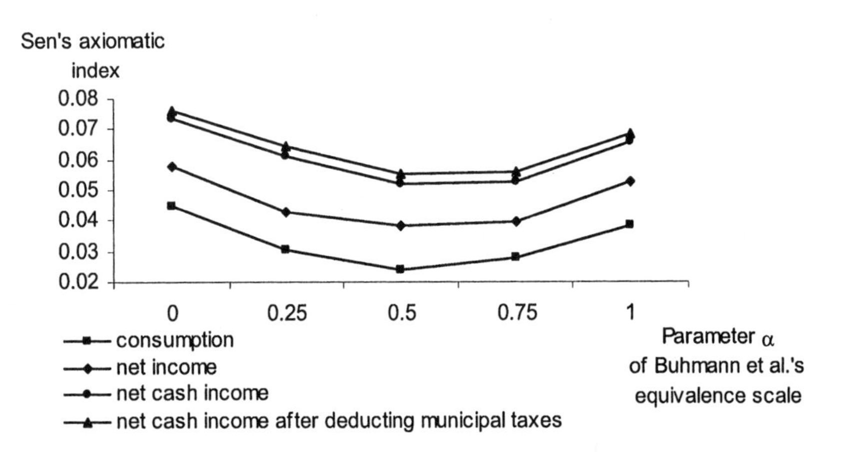

Fig. 4.2. Impact of parameter α of *BES*(α) on Sen's axiomatic index when employing *EWI*. The same U-shaped pattern also appears when employing either the other four poverty indices or *EWH*. Note that the ranking of the various economic well-being variables corresponds to the order stressed in Sect. 4.2.

4.4 Choice of definition and identification of the poor

In the previous section I analyzed whether the selection of a weighting procedure, equivalence scale and economic well-being variable affect the scope of poverty; here I analyze whether these choices systematically influence identification of the poor population. Specifically, I show how a household that is classified as poor given one combination of weighting procedure, equivalence scale and economic well-being variable is not considered poor when using another combination of these same factors. Previous studies have shown that the population defined as poor on the basis of income is not identical to the population so identified on the basis of consumption. This section, however, presents a more systematic and broader analysis that compares not only economic well-being variables but also weighting procedures and equivalence scales.

4.4.1 Looking at the poorest quintile

Table 4.2 indicates the distribution of households by net cash income and consumption deciles, assuming the equivalence scale is *BES*(α = 0.5) and the weighting procedure *EWH*. The table shows, for example, that among households belonging to the poorest quintile defined on the basis of net cash income, only 10.6% also belong to the poorest quintile defined in terms of consumption. This corresponds to an *agreement ratio* of 53.1% (i.e., 10.6% out of 20%). About 12% (2.4% out of 20%) of the households that belong to the poorest quintile of the net cash income exhibit a consumption level that places them above the median of the net cash income.

Table 4.2. Classification of households by net cash income and consumption deciles

Consumption deciles	Net cash income deciles										Total
	1	2	3	4	5	6	7	8	9	10	
1	3.3	2.9	1.9	0.9	0.4	0.4	0.1	0.0	---	---	10
2	2.2	2.1	1.8	1.6	1.0	0.6	0.4	0.2	0.1	0.0	10
3	1.4	1.6	1.1	1.8	1.4	1.2	0.8	0.4	0.2	0.0	10
4	1.2	1.2	1.6	1.4	1.5	1.4	0.7	0.6	0.3	0.1	10
5	0.8	0.8	1.1	1.4	1.1	1.4	1.4	1.0	0.7	0.3	10
6	0.4	0.4	1.0	1.0	1.5	1.6	1.5	1.2	1.0	0.5	10
7	0.3	0.4	0.6	0.7	1.2	1.3	1.8	1.6	1.4	0.7	10
8	0.1	0.1	0.4	0.6	1.0	1.0	1.6	2.1	1.8	1.3	10
9	0.2	0.2	0.3	0.5	0.6	0.8	1.0	1.7	2.2	2.6	10
10	0.1	0.2	0.1	0.2	0.4	0.4	0.7	1.1	2.3	4.4	10
Total	10	10	10	10	10	10	10	10	10	10	100

Note: The numbers represent percentages of the total population of 1,504,928 households, computed by employing *BES*($\alpha = 0.5$) and *EWH* for both income and consumption.

One can also observe that 7% (0.7% out of 10%) of the households that belong to the first (poorest) decile of the net cash income are also found in the seventh (or higher) decile of the consumption. At the same time only 1% (0.1% out of 10%) of those found in the first (poorest) decile of the consumption distribution are also found in the seventh (or higher) decile of the net cash income. Similar results are obtained when comparing the distribution of net income or net cash income after deduction of local taxes with that of consumption. It appears that such a pattern is not unique to Israel. McGregor and Borooah (1992), for example, show similar results using 1985 British household expenditures.

Table 4.3 summarizes the results of the agreement ratios between the lowest quintile of each type of income and the lowest quintile of consumption for each of the selected weighting procedures and equivalence scales.

Table 4.3 can be used to study the impact of weighting procedure on the agreement ratio. As seen from the table, comparing the distribution of net income with that of consumption yields a smaller agreement ratio when equal weight is given to individuals rather than to households. The conclusions are less clear when the distribution of consumption is compared with that of net cash income or of net cash income after deduction of local taxes.

Table 4.3. Agreement ratios for the first quintile

Equivalence scale	Consumption with net cash income after deduction of local taxes		Consumption with net cash income		Consumption with net income	
	EWI	EWH	EWI	EWI	EWI	EWI
BES(α = 0)	60.5%	62.8%	61.8%	63.5%	64.2%	67.5%
BES(α = 0.25)	55.8%	56.5%	56.2%	57.3%	58.9%	63.6%
BES(α = 0.5)	52.7%	51.3%	53.7%	53.1%	57.9%	60.8%
BES(α = 0.75)	53.0%	51.7%	53.9%	53.4%	59.6%	61.0%
BES(α = 1)	59.6%	55.8%	60.9%	56.4%	63.3%	64.1%
NII-ES	52.3%	51.3%	53.4%	52.8%	58.1%	60.8%

Next, the impact of the equivalence scale on the agreement ratio is considered. Fig. 4.3 shows that a U-shaped curve is obtained when looking at the relationship between the agreement ratio and the parameter α of *BES*(α). This result holds irrespective of which of the three income concepts is compared with consumption. Fig. 4.3 also shows that the agreement ratio is highest when a comparison is made between consumption and net income. It is somewhat lower when consumption is compared with net cash income and even lower when consumption is compared with net cash income after deduction of local taxes. Although Fig. 4.3 is based on *EWH* as the weighting procedure, similar results are obtained when *EWI* is employed as the weighting procedure.

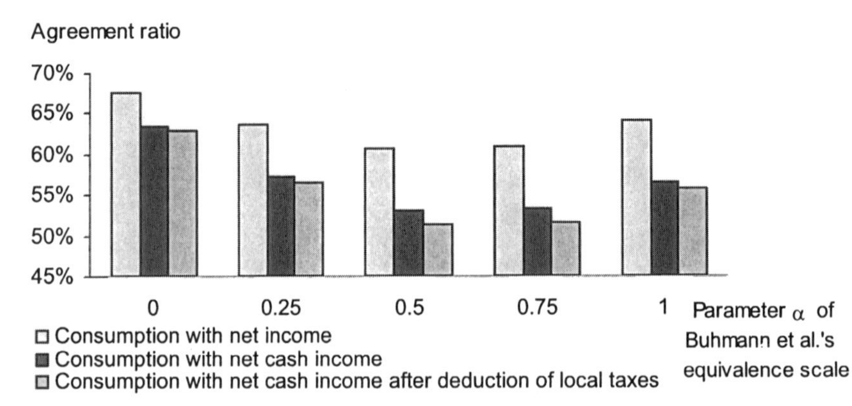

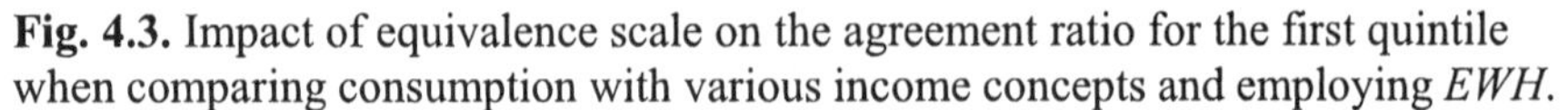

Fig. 4.3. Impact of equivalence scale on the agreement ratio for the first quintile when comparing consumption with various income concepts and employing *EWH*.

4.4.2 The population defined as poor

We now turn to the agreement ratio when comparing the population defined as poor according to each of the possible combinations of weighting procedures, equivalence scales and economic well-being variables. The agreement ratio in the current case is defined as $\#(A \cap B)/\#(A)$, where $\#(A \cap B)$ represents the number of households classified as poor in cases A and B, with $\#(A)$ representing the number of households classified as poor in case A.

Table 4.4 illustrates this type of comparison for our six equivalence scales and two weighting procedures when the distribution of consumption is compared with itself (in such a comparison, all the numbers along the diagonal are equal to 100 by definition). It should be noted that as one moves away from the diagonal (to the right or left, upward or downward), the agreement ratio decreases whereas the agreement ratio increases as the values of α in the two compared distributions converge.

The decrease in the agreement ratio observed in Table 4.4 as the distance from the diagonal increases is also observed when employing each of our three income types. The same pattern is observed when the distribution of net cash income after deduction of local taxes is compared with the distribution of net cash income. Obviously, in that case, the numbers on the diagonal will generally be different from 100. About 96% to 100% of the population defined as poor according to the distribution of net cash income are also classified as poor according to the distribution of net cash income after deduction of local taxes provided that the same equivalence scale and weighting procedure is used in the comparison. Similarly, 93% to 98% of those classified as poor on the basis of net cash income after deduction of local taxes are also poor on the basis of net cash income.

Table 4.5 presents other interesting results regarding the agreement ratios. The table shows that the smallest agreement ratio is obtained when one compares distributions derived from various equivalence scales for a selected economic well-being variable and a given weighting procedure. The data indicate that under these conditions, the minimal agreement ratio varies between 19% and 41% and always corresponds to the case where one distribution is derived on the basis $BES(\alpha = 1)$ and the other distribution corresponding to the case with $BES(\alpha = 0)$. In other words, 59% to 81% of the population defined as poor when one of the equivalence scales is used (parameter α equals to 0 or 1) is not considered poor when the alternative value of the parameter α is selected (1 or 0). Note also that when employing *EWH,* the minimal agreement ratio is higher than in the case using *EWI.*

Table 4.4. Agreement ratios when comparing consumption distribution with itself by weighting procedure

		Consumption (B)											
		Equal weight given to individuals						Equal weight given to households					
		Equivalence scale						Equivalence scale					
		0	0.25	0.5	0.75	1	NII	0	0.25	0.5	0.75	1	NII
Consumption (A): Equal weight given to households, Equivalence scale	NII	73	75	77	75	65	77	60	70	81	97	83	100
	1	43	45	46	(53)	55	52	33	41	49	71	100	64
	0.75	66	68	70	70	66	72	53	64	74	100	87	91
	0.5	91	93	91	63	48	68	78	89	100	78	64	81
	0.25	99	100	67	43	31	48	91	100	73	55	44	57
	0	100	81	46	27	18	31	100	72	50	36	27	38
Consumption (A): Equal weight given to individuals, Equivalence scale	NII	77	82	88	93	74	100	63	76	89	100	88	100
	1	54	59	65	83	100	78	39	53	67	97	100	90
	0.75	72	76	83	100	80	95	57	71	84	100	93	100
	0.5	91	95	100	68	52	74	79	90	100	82	66	85
	0.25	98	100	59	39	29	43	87	84	65	50	40	52
	0	100	67	39	25	19	28	74	58	43	33	27	35

Note: The numbers 0, 0.25, 0.5, 0.75 and 1 in the equivalence scale title refer to the parameter α of $BES(\alpha)$. NII refers to the *NII-ES*. The circled number (53) indicates that 53% of the households that are defined as poor according to consumption distribution A when the equivalence scale is $BES(\alpha = 1)$ and the weighting procedure is *EWH* are also classified as poor according the consumption distribution B when the equivalence scale is $BES(\alpha = 0.75)$ and weighting procedure is *EWI*.

When the same concept of minimal agreement ratio is employed with distributions of different economic well-being variables, the minimal agreement ratio is even smaller – and may reach as low as 11%. Four cases corresponding to this minimal value occur when consumption is compared with either net cash income or net cash income after deduction of local taxes. In all these cases, the comparison corresponds to the case where α equals 0 in the consumption distribution and 1 in the income distribution.

Table 4.4. Minimal agreement ratio when comparing different equivalence scales using the same economic well-being variable

Economic well-being variable	EWI	EWH
Consumption	19%	27%
Net income	25%	39%
Net cash income	27%	39%
Net cash income after deduction of local taxes	29%	41%

4.4.3 Identifying the poor according to more than two distributions

It may also be of interest to ascertain what part of the population is considered poor according to at least one of the 48 unidimensional approaches considered, to all the approaches or, alternatively, to a subset of those approaches. As there are $(2^{48} - 1)$ possible subsets of different combinations of distributions, it is impractical to consider them all. Table 4.6 summaries the scope of poverty (measure by *H*) using an assortment of subsets.

Table 4.5. The scope of poverty (*H*) according to assorted subsets of the 48 distributions

Subsets (No. of different subsets)	Any single distribution	At least one distribution	All distributions
With one distribution (48)	8.02 – 24.26	40.12	0.60
With two distributions (1,128)	1.82 – 23.94	35.52	0.60
With three distributions (17,296)	1.12 – 20.49	33.34	0.60
With four distributions (194,500)	0.70 – 17.56	31.62	0.60
With 44 distributions (194,500)	0.60 – 0.97	1.55	0.60
With 45 distributions (17,296)	0.60 – 0.87	1.35	0.60
With 46 distributions (1,128)	0.60 – 0.76	1.09	0.60
With 47 distributions (48)	0.60 – 0.70	0.84	0.60
With one consumption distrib. (12)	8.02 – 20.50	27.25	2.54
With one net cash income distrib. (12)	11.92 – 23.86	30.57	5.68
With one net income distrib. (12)	9.43 – 23.62	27.95	4.47
With one net cash income after deduction of local taxes distribution (12)	12.35 – 24.26	30.67	5.92

Note: The numbers are expressed as the percentage of poor persons to total population (4,984,871 persons).

As Table 4.6 shows, only about 0.6% of the population is classified as poor in all the distributions (compared to from 8.02% to 24.26% defined as poor according to any one distribution). A low income as well as low consumption level characterize this population whatever the equivalence scale used. On the other hand, 59.88% of the population is never defined as poor no matter which distribution is analyzed. If considered from a different perspective, 40.12% of the population is poor according to at least one distribution. When we consider the population that is defined as poor according to any two different distributions (a total of 1,128 possible subsets), we find that about 1.82% to 23.94% of the population fall into that category. A total of 35.52% of the population is poor according to at least one of these subsets of two distributions. Table 4.6 also presents the corresponding results when we consider subsets of 3, 4 and 44 to 47 distributions. The table thus enables us to observe the convergence of the different distributions.

Moreover, Table 4.6 enables us to consider this convergence when employing the same economic well-being variable (the last 4 rows of the table). For instance, only 4.47% of the population is defined as poor according to their net income level, irrespective of the equivalence scale and the weighting procedure.

In fact, in order to define, if only approximately, who is poor according to the entire range of distributions, it is sufficient to look at the distributions of net income, net cash income and consumption employing $BES(\alpha = 0)$ and $BES\ (\alpha = 1)$ together with the two weighting procedures. By limiting the search to these 12 distributions, we find that 0.62% of the total population is poor under any of these 12 distributions represented, whereas 38.74% of the population is poor according to at least one of the distributions.

4.5 Concluding comments

This study shows that the choice of the economic well-being variable, equivalence scale and weighting procedure has an impact on the values obtained with various poverty indices in addition to how poverty is identified.

The main findings can be summarized as follows: Poverty indices are higher when equal weight is given to each household rather than when each household is weighted by the number of its members. The level of poverty also depends on the economic well-being variable selected to measure it. Poverty is highest when the economic well-being variable adopted is net cash income after deduction of local taxes. Poverty is lowest when it is measured on the basis of consumption. Since poverty is higher when using net cash income after deduction of local taxes as opposed to net cash income, one may conclude that local taxes tend to increase poverty. Similarly, since poverty derived on the basis of net cash income is higher than when it is derived from net income, one may infer that imputed income and income in kind tend to reduce poverty. The analysis also shows that the relationship between poverty and the value of the equivalence elasticity defined by Buhmann et al. (1988) is U-shaped.

An examination of the population at the bottom of the income and consumption distributions indicates that only 51.3% to 67.5% of the population belonging to the first quintile according to one of the income distributions will also be found in the first quintile of the consumption distribution. Moreover, here again we find a U-shaped relationship between the agreement ratio with respect to the first quintile and the equivalence elasticity defined by Buhmann et al. (1988). The highest level of agreement (regarding the first quintile) was found when comparing net income and consumption, whereas the lowest was found when comparing net cash income after deduction of local taxes and consumption. The agreement ratio regarding the poor population was likewise found to be directly related to the kind of equivalence scale adopted. Local taxes were not, however, found to have a significant impact on the identification of the poor population since 96% to 100% of those who are poor according to net cash income are also poor according to net cash income after deduction of local taxes.

When looking at the 48 distributions at the basis of the analysis, we discover that only 0.6% to 0.62% of the population is defined as poor by all 48 distributions based on the 1997 data. This population represents the "hard core" of Israel's poor. Nevertheless, between 40.12% and 42.45% of Israel's population is defined as poor according to at least one of the distributions.

References

Abul Naga RH (1994) Identifying the Poor: A Multiple Indicator Approach, Distributional Analysis Research Programme (DARP) Discussion Paper 9, London: London School of Economics

Atkinson AB (1987) On the Measurement of Poverty. Econometrica 55:749-764

Atkinson AB (1991) Comparing Poverty Rates Internationally: Lessons from Recent Studies in Developed Countries. The World Bank Economic Review 5:3-22

Atkinson AB (1992) Measuring Poverty and Differences in Family Composition. Economica 59:1-16

Banks J, Johnson P (1994) Equivalence Scale Relativities Revisited. Economic Journal 104:883-890

Buhmann B, Rainwater L, Schmaus G, Smeeding TM (1988) Equivalence Scales, Well-being, Inequality, and Poverty: Sensitivity Estimates across Ten Countries Using the Luxembourg Income Study (LIS) Data Base. Review of Income and Wealth 34:115-145

Burkhauser RV, Smeeding TM, Merz J (1996) Relative Inequality and Poverty in Germany and the United States Using Alternative Equivalence Scales. Review of Income and Wealth 42:381-400

Chakravarty SR (1983) Ethically Flexible Measures of Poverty. Canadian Journal of Economics 16:74-85

Clark S, Hemming R, Ulph D (1981) On Indices for the Measurement of Poverty. Economic Journal 9:515-526

Coulter FAE, Cowell FA, Jenkins SP (1992) Equivalence Scale Relativities and the Extent of Inequality and Poverty. Economic Journal 102:1067-1082

Cowell FA (1984) The Structure of American Income Inequality. Review of Income and Wealth 30:351-375

Cowell FA, Mercader-Prats M (1999) Equivalence Scale and Inequality. In Silber J (ed) Handbook on Income Inequality Measurement. Kluwer Academic Publishers, Boston, pp 405-436

Danziger S, Taussig MK (1979) The Income Unit and the Anatomy of Income Distribution. Review of Income and Wealth 25:365-375

Davidson R, Duclos JY (2000) Statistical Inference for Stochastic Dominance and for the Measurement of Poverty and Inequality. Econometrica 68:1435-1464

Deaton A, Muellbauer J (1980) Economics and Consumer Behavior. Cambridge University Press, Cambridge

Ebert U (1999) Using Equivalence Income of Equivalent Adults to Rank Income Distributions. Social Choice and Welfare 16:233-258

Foster JE (1984) On Economic Poverty: A Survey of Aggregate Measures. In Basmann RL, Rhodes GFJ (eds) Advances in Econometrics 3. JAI Press, Connecticut, pp 215-251

Foster JE, Greer J, Thorbecke E (1984) A Class of Decomposable Poverty Measures. Econometrica 52:761-766

Glewwe P, Van Der Gaag J (1990) Identifying the Poor in Developing Countries: Do Different Definitions Matter?. World Development 18:803-814

Hagenaars AJM (1987) A Class of Poverty Indices. International Economic Review 28:583- 607

Hagenaars AJM, De Vos K (1988) The Definition and Measurement of Poverty. The Journal of Human Resources 23:211-221

Jenkins SP, Cowell FA (1994) Parametric Equivalence Scales and Scale Relativities. Economic Journal 104:891-900

Jenkins SP, Lambert P (1997) Three 'I's of Poverty Curves, with an Analysis of UK Poverty Trends. Oxford Economic Papers 49:317-327

Jenkins SP, Lambert P (1998a) Ranking Poverty Gap Distributions: Further TIPs for Poverty Analysis. Research on Economic Inequality 8:31-38

Jenkins SP, Lambert P (1998b) Three 'I's of Poverty Curves and Poverty Dominance: TIPs for Poverty Analysis. Research on Economic Inequality 8:39-56

McGregor PL, Borooah VK (1992) Is Low Spending or Low Income a Better Indicator of Whether or not a Household is Poor: Some Results From the 1985 Family Expenditure Survey. Journal of Social Policy 21:53-69

McGregor PL, Nachane D (1995) Identifying the Poor: A Comparison of Income and Expenditure Indicators Using the 1985 Family Expenditure Survey. Oxford Bulletin of Economics and Statistics 57:119-128

Menirav E (2002) On the Effect of Measurement on Poverty. Economics Quarterly 49: 321-354 [Hebrew]

Mercader-Prats M (1998) Identifying Low Standards of Living: Evidence from Spain. Research on Economic Inequality 8: 155-173

Ravallion M (1994) Poverty Comparisons. Fundamentals in Pure and Applied Economics 56. Harwood Academic Publishers, Chur, Switzerland

Sen A (1976) Poverty: An Ordinal Approach to Measurement. Econometrica 44:219-231

Sen A (1979) Issues in the Measurement of Poverty. The Scandinavian Journal of Economics 81:285-307

Shorrocks AF (1995) Revisiting the Sen Poverty Index. Econometrica 63:1225-1230

Slesnick DT (1993) Gaining Ground: Poverty in the Postwar United States. Journal of Political Economy 101:1-38

Smeeding TM, Saunders P, Coder J, Jenkins SP, Fritzell J, Hagenaars AJM, Hauser R, Wolfson M (1993) Poverty, Inequality, and Family Living Standards Impacts Across Seven Nations: The Effects of Noncash Subsidies for Health, Education and Housing. Review of Income and Wealth 39:229-256

Thon D (1979) On Measuring Poverty. Review of Income and Wealth 25:429-440

Watts HW (1968) An Economic Definition of Poverty. In Moynihan DP (ed) On Understanding Poverty: Perspective from the Social Sciences. Basic Books, New York, pp 316-329

Zheng B (1997) Aggregate Poverty Measures. Journal of Economic Surveys 11:123-162

5 Capability Approach and Fuzzy Set Theory: Description, Aggregation and Inference Issues

Enrica Chiappero Martinetti

Dipartimento Economia Pubblica e Territoriale, University of Pavia

5.1 Introduction

The capability approach, initially elaborated by Amartya Sen, represents an important point of reference in the research field of poverty and well-being analysis. Its success is due to the fact that it is more than a mere multidimensional framework; it provides a broad, rich, and intrinsically complex perspective for describing the multifaceted nature of poverty and of well-being, for understanding their causes and effects, and for investigating interrelated layers of analysis that are often neglected or inadequately discussed.

At a theoretical level, these aspects represent points of strength within the capability approach. However, they can also generate methodological and technical matters that are not easily resolved. Nevertheless, since its formulation in the mid-1980s, a significant amount of empirical applications based on, or inspired by, the capability approach have been produced. Various statistical tools have also been tested, including fuzzy sets methodology, which has been applied with interesting results[1]. In all these empirical applications of the capability approach, fuzzy set theory has been used mainly in order to depict deprivation and well-being indicators in a gradual, rather than dichotomous, manner; some of them have also focused attention on aggregative issues (Chiappero-Martinetti 2000; Baliamoune 2004) while a few (Chiappero-Martinetti 1996) have addressed the inference problem through the use of fuzzy logic.

This Chapter aims to take a further step forward in this direction by trying to show how fuzzy methodologies can be powerful tools for implementing three different, and often consequential or complementary, exercises that characterize every multidimensional assessment of poverty and well-being: i) the *description phase,* ii) the *aggregation phase* and iii) the *inference phase*.

---

[1] See, in particular, Chiappero-Martinetti (1994, 1996, 2000, 2005), Lelli (2001), Clark and Qizilbash (2002), Baliamoune (2004).

What I call the *representation or description* phase is a sort of extension of the identification step in the traditional approach to poverty analysis: in a fuzzy environment, the conventional "hard" threshold, which determines an unambiguous distinction between "poor" and "not poor", is substituted by a "soft" threshold that depicts an intermediate, gradual representation between acceptable and unacceptable living conditions, or adequate and inadequate levels of well-being, without establishing a single abrupt cut-off point. This also makes it possible to conceive the analysis of poverty and well-being not as two separate and distinct exercises, as has traditionally been the case, but as two intertwined facets of a broader conceptual framework, such as the capability approach. A further advantage of fuzzy set theory is that this sort of gradual membership can be used in order to represent not only quantitative (continuous or discrete) variables, but also qualitative indicators, linguistic attributes, and hedges or qualifiers[2], that are typically included in questionnaires or data sets on subjective or objective well-being, and are undoubtedly central from a capability perspective.

The *aggregation phase* is conceived here not in terms of an aggregative exercise across individuals (for which an extensive literature on poverty measurement already exists), but mainly as aggregation across dimensions or domains of well-being for each unit of analysis, whether they are individuals or households. This is a crucial step in every multidimensional assessment that does not require dealing with technical aspects of the aggregative process alone. Combining elementary dimensions of well-being in order to achieve an overall assessment requires making what are largely value judgments and normative implications in the underlying theoretical concepts explicit. The availability of a variety of fuzzy sets operators makes it possible to better adapt and calibrate the aggregation phase to the theoretical framework.

Finally, the *inference* phase refers to the possibility of applying fuzzy logic rules and fuzzy approximate reasoning in order to infer a logical conclusion (i.e. the existence of a capability deprivation, the lack of freedom

---

[2] In brief, linguistic variables are words or sentences expressed in natural language. Age, for instance, is a quantitative variable when it is expressed in terms of years, but becomes a linguistic variable when referred to with a (fuzzy) predicate such as "old" or "young". A linguistic hedge or fuzzy quantifier modifies the meaning of a predicate or, more generally, of a fuzzy set: *very*, *close to*, *quite*, *fairly*, are all examples of hedges. In the same fashion as adverbs and adjectives in language, these qualifiers change the shape of fuzzy sets. For instance, by applying the hedge *very* to the linguistic variable or fuzzy set of "young people", we obtain a different fuzzy set, and thus a different representation of the corresponding membership function.

to achieve, or exposure to the risk of poverty) starting from premises that are known or assumed to be true.

These three steps will be discussed taking the capability approach as the main point of reference; however, the methodological and empirical issues and suggestions presented here can be adapted equally well to any other multidimensional framework for the analysis of well-being and poverty.

This Chapter is organized as follows: Sect. 5.2 introduces the main features and methodological requirements of the capability approach. Sects. 5.3, 5.4 and 5.5 discuss the three phases that characterize a multidimensional well-being assessment from a capability perspective (respectively, description, aggregation and inference). Sect. 5.6 concludes.

5.2 Brief remarks on distinctive features of the capability approach

The capability approach characterizes individual well-being in terms of what a person is actually able to do or to be[3]. Functionings and capabilities are two of the key concepts within this theoretical framework, and identify two distinct – though strictly interrelated - spaces that can be used to understand and assess well-being.

Basically, functionings are the valuable achievements, actions, and activities that determine individual well-being: they are, in Sen's language, "the various things a person may value doing or being" (Sen 1992). These include achievements such as being nourished, being healthy, being educated, taking part in social life and in political decisions, just to mention some of the most common.

Capabilities are combinations of beings and doings that a person can achieve and reflect the real set of options that a person has to achieve what she or he values. While functionings constitute a person's achieved well-being, capabilities represent the real opportunities for a person to achieve well-being, and thus include the freedom of choice.

Achievements – i.e., the functionings set – as well as the possibility to achieve functionings – i.e., the capability set – for a given person are both influenced by internal or personal factors (such as age, gender, health, and disability) as well as by external or "environmental" circumstances (including household structure, socio-economic context, cultural and social norms, institutions, the natural environment, and so forth). The process of

[3] There has been an exponential growth in the literature on the capability approach over the past ten years. For a recent theoretical survey, see Robeyns (2005) and the references therein.

conversion of available resources (e.g. market and non-market goods, commodities and services, income, etc.) into well-being is closely related to, and dependent on, these individual and environmental features.

Formally, (Sen 1985; Kuklys 2005) the individual capability set Q_i is the space of potential functionings, and can be expressed in the following way:

$$Q_i(X_i)=\{\mathbf{b}_i \mid \mathbf{b}_i = f_i(\mathrm{x}_i) \mid (\mathbf{h}_i, \mathbf{e}_i)\} \quad (5.1)$$

for some $f_i(\cdot) \in \mathrm{F}_i$ and some $x_i \in \mathrm{X}_i$, where $\mathbf{b}$ is a vector of functionings, f_i is a conversion function, and $\mathbf{h}_i$ and $\mathbf{e}_i$ are respectively vectors of personal factors and external or environmental factors which influence the rate of conversion of individual resources (x_i) to a given functioning (b_i)[4].

This thumbnail sketch makes the way in which the capability approach offers a genuine multivariate perspective evident. First of all, two distinct evaluative spaces – the capability space and the functioning space – can be taken into account, distinguishing what individuals *can do* from what they *actually do*[5]. Secondly, both evaluative spaces can be seen as multidimensional spaces themselves: functionings such as education, health, participation, and housing are typically assessed using a plurality of qualitative and quantitative variables. Thirdly, the way in which the capability approach conceives human diversity is intrinsically pluralistic: it acknowledges how the high degree of heterogeneity in personal features such as sex, age, and physical and psychological conditions makes each person substantially different from the next, generating deep interpersonal variations in the conversion of resources into functionings and capabilities. Fourthly, two hypothetically identical people living within different households, different socioeconomic contexts, and different cultural and natural environments will generally have different sets of opportunities and different living conditions. Finally, what makes the capability approach an intrinsically complex (and not only multidimensional) approach is the attention it focuses on investigating the links among the aforementioned layers of analysis[6].

---

[4] As pointed out by Kuklys (2005) while X_{ii} represents the resource constraint, $\mathbf{h}_i$ and $\mathbf{e}_i$ can be interpreted as non-monetary constraints.

[5] Even if, typically, these two sets have been seen as alternative, elsewhere I have argued why it can be important to include both spaces in a well-being assessment, rather than to choose between them (Chiappero-Martinetti 1996).

[6] What are the differences and the relationships between capability and functioning sets, what kind of practical advantages and technical difficulties are associated with the decision to focus attention on one space or another, how can pervasive human diversities and heterogeneous contexts positively or negatively influence our achievements, and our possibility to achieve our overall freedom, via the conversion factors? Traditionally, these are the most common questions that the capability literature deals with.

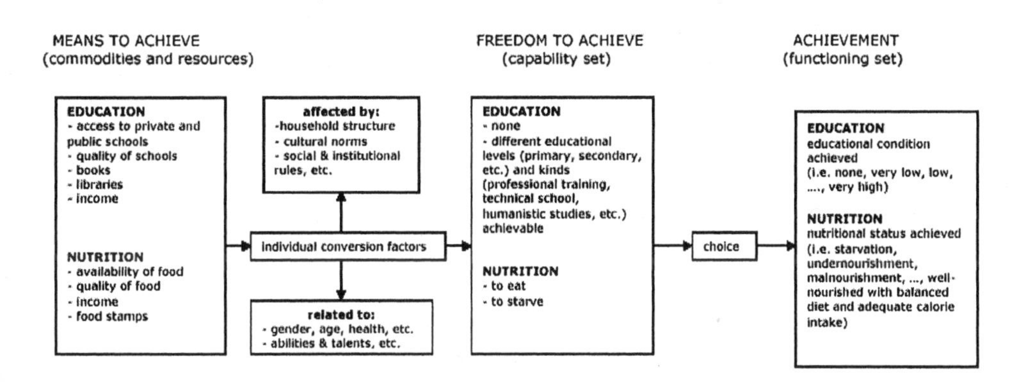

Fig. 5.1. A simplified representation of the capability approach

Fig. 5.1. offers a diagrammatic representation of the capability approach based on a straightforward example in which only two evaluative spaces – education and nutrition – are considered[7].

What are the most challenging methodological aspects in terms of the operationalization of the capability approach? Firstly, we need to be able to describe human poverty and individual well-being in all its multifaceted and gradual aspects. Heterogeneity, and differences in terms of achievements, the opportunity to achieve, and interpersonal features and contexts should be reflected and preserved in the description of human well-being.

This necessitates, on the one hand, dealing with a plurality of qualitative and quantitative indicators related to both evaluative spaces (i.e. capability and functionings sets) as well as to the individual and environmental characteristics; on the other hand, recognizing that capability deprivation, individual well-being, or happiness should be depicted using different degrees of intensity and nuances as opposed to "all or nothing" conditions. Secondly, we might be interested in summarizing indicators and dimensions into an overall measure of individual well-being, and thus we would need to have at our disposal aggregative criteria that allow us to preserve to as great an extent as possible elementary information, and to reflect in an appropriate way the relative importance that we want to assign to each single dimension.

Finally, in order to investigate patterns of causation among indicators and/or dimensions, as well as to assess well-being and (unobservable) capabilities starting from (observable) statistical data on achieved beings and doings, we need to use some theory of inference, that is, we need to derive

[7] This figure is a simplified and revised version of a diagram suggested by Robeyns (2005).

logical conclusions from premises that are known or from factual knowledge or evidence. As I will argue over the next sections of this Chapter, fuzzy set theory and fuzzy logic seem to be powerful tools for undertaking all three of three exercises.

5.3 Describing multidimensional poverty and well-being through fuzzy membership functions

In this section I will try to illustrate how fuzzy methodology can be applied to describe and assess achieved functionings $\mathbf{b}_i$. I will do so by making use of the bi-dimensional example mentioned above and describing a sort of step-by-step procedure. Our attention is focused here on the right block of the diagram in Fig. 5.1, i.e. on the final part of the well-being process, which refers to the space of achieved functionings.

The first step that must be undertaken requires identifying the relevant evaluative domains and specifying the corresponding fuzzy sets appropriately. From a capability perspective, this step requires not only identifying the valuable *beings* and *doings* on which the empirical analysis will be based[8], but also trying to establish a close correspondence between the theoretical concept that we want to describe – i.e. functioning achievement – and its formal representation in terms of fuzzy sets. As Ragin (2000) points out, specifying fuzzy sets is different and more complex than specifying variables. At a technical level, it requires establishing the number and nature of the variables and identifying a suitable and accurate range of values, scales, and modalities for each variable. It may be necessary to underline what are considered crucial aspects and distinctive features of our investigation, or to pay special attention to a given range of values or modalities of the chosen variables. In some cases only a limited number of values or modalities are possible or necessary for representing a theoretical concept, while in others a fine-grained distinction is both conceptually relevant and empirically possible. For example, with reference to the functioning "*being nourished*", concepts such as "starvation", "malnourishment", "undernourishment", or "unbalanced diet" not only convey different semantic meanings but also express different conditions in terms of nutritional status when related to different contexts (e.g. a poor country or

[8] It is not the aim of this Chapter to discuss here the controversial question in the capability debate on whether a list of valuable functionings should be predetermined or not and how, when and who should select and determine this list. On this issue, that mainly pertains to the normative, value-judgments sphere, see Sen (2004), Nussbaum (2003), Alkire (2002) and Robeyns (2003).

an advanced industrial society, a rural or an urban area) or specific groups of people (i.e. babies, girls, adult males). In order to capture the concepts of "malnourishment" or "unbalanced diet", it may not be sufficient to establish different quantities or thresholds for given variables (i.e. food intake, amount of calories); both quantitative and qualitative variables may need to be taken into account, and the type and specification of these variables may well be different in an affluent society, where a high percentage of fats in daily diet or an excess calorie intake may be symptoms of an "unbalanced diet", and in a poor country, where a vitamin or protein deficiency and/or deficiency in the daily calorie intake will be symptomatic of a condition of malnourishment.

Table 5.1 shows an example of verbal labels or qualifiers that we could apply to two common functionings, i.e. "*being educated*" and "*being nourished*"[9]; and, in parentheses, a possible specification of these "*beings*" ranked in an increasing order[10]. Some clarification might be helpful here. Firstly, the use of a common qualitative range of attributes (from none, very low, low, and so forth to very good), such as in the above example, even if not strictly necessary, makes it possible to generalize the description exercise while maintaining the specificity of each dimension (i.e., the same qualitative labels are used for both dimensions but with an *ad hoc* correspondence according to the given well-being dimension being examined, i.e. the meaning of the label "good" is different when referred to education than it is when referred to nutrition).

Secondly, to each label or attribute it is possible to associate qualitative correspondences, quantitative distinctions or a mix of both (i.e. "balanced diet" or "minimum calorie intake corresponding to 2000 calories per day" or "minimum calorie intake and balance diet achieved", as in Table 5.1), embracing both differences in kind and differences in degree[11].

Thirdly, and closely related to the previous point, differences in kind (as qualitatively expressed by categorical distinctions) also qualify differences

---

[9] In order to simplify, attention will be restricted here to the simplest bi-dimensional case with a single variable for each functioning. However, this example could easily be extended to a more relevant and realistic case in which a plurality of indicators is taken into consideration for a plurality of well-being domains. In such a case, the set of elementary indicators to a given dimension, as well as the multiplicity of dimensions, could be combined according to one of the aggregation procedures that will be discussed in Sect. 5.4.

[10] Other possible labels quite common in qualitative analysis include qualifiers that rank from "completely true" to "not at all true", or from "fully in" to "fully out", and so on.

[11] See Ragin (2000) on the dual concept of diversity in terms of categorical diversity and differences in degree.

in degree (as quantitatively expressed by membership degree). For instance, in terms of everyday language, it is easy to recognize that adjectives such as "inadequate", "inappropriate" or "insufficient" when referred to educational or nutritional levels communicate a different quantitative as well as qualitative degree of deprivation, underlining how achievements are more a matter of degree than they are "all or nothing" conditions. Similarly, functionings such as "being *adequately* nourished", "being *well*-sheltered" or "being in *good* health" (Sen 1992) suggest a complex and gradual living condition that can hardly be captured by drawing an abrupt cut-off point or threshold. Finally, the same labels can be adapted to represent different conditions depending on individual status and specific context, e.g., a low or inadequate level of nutrition will be quantitatively and qualitatively different for a child, a girl or a pregnant woman, a worker living in a rural area or a white-collar employee, according to climatic conditions, tradition, social customs, etc.

Table 5.1. An example of gradually achieved functionings

"Being educated"	"Being nourished"
None (illiterate)	None (totally insufficient for survival; starvation)
Very low (ability to write and read only, but no formal education)	Very low (sporadic access to food and very serious under-nourishment)
Low (attendance at the lowest level of formal education, but with discontinuity and/or dropout)	Low (malnourishment; inadequate diet)
Sufficient (lower formal level of education achieved)	Sufficient (minimum calorie intake almost achieved, but diet not fully balanced)
Quite good (attendance in secondary school, but with discontinuity and/or dropout)	Quite good (minimum calorie intake achieved, but diet not fully balanced)
Good (secondary school achieved)	Good (minimum calorie intake achieved, and diet quite balanced)
Very Good (access to/achievement of the highest level of secondary school education)	Very good (nutritional level and balanced diet achieved)

At a first glance there would not seem to be any particular difference between this (fuzzy) representation and an ordinal scale that yields to a ranking order among categories. There is, however, meaningful dissimilarity, also in this preliminary descriptive phase: while an ordinal scale is a mere ranking of categories, fuzzy methodology translates these ordinal ranks into fuzzy membership scores or degrees that are capable of reflecting the content of the ordinal categories in line with our conceptual understanding of the phenomenon that we want to describe.

This leads us to the second step of the description phase, i.e. how to assign membership degrees or scores and to calibrate appropriate membership functions. Again, this step is neither automatic nor univocal as it would be in the case of an ordinal scale. Different methods can be adopted for constructing membership functions.

First of all, they can be chosen by the investigator arbitrarily, according to her or his common sense and experience, or the value judgements underlying the theoretical concept that she or he wishes to describe. For instance, a simple decreasing or increasing linear membership function can be adequate in order to depict variables or concepts (i.e. functionings) distributed along a linear continuum between 0 and 1 (inclusive), where any value is proportional to its distance in the value axis (i.e. with reference to Table 5.1, the distance between "none" and "very low" is the same that separates the modalities "very low" and "low" and so on). Trapezoidal-shaped membership functions make it possible to preserve linearity and at the same time to incorporate minimum and/or maximum thresholds: this can be helpful in adapting functionings to different realities or circumstances. For instance, with regard to the achieved functioning of "*being educated*", a middle-income country might associate a condition of full deprivation with, say, the lowest three ranks in Table 5.1, while a less-developed country might associate a condition of full achievement with the higher two ranks[12].

Nonlinear membership functions – such as sigmoid or logistic curves – can make it possible to "fine-tune" the representation and require identifying not only the two extreme 0 and 1 membership values, but also the flex or crossover point associated to a membership degree equal to 0.5, according to criteria established by the investigator. Other nonlinear functions, such as Gaussian or exponential curves or irregularly shaped functions can be equally applied.

Second, membership grade functions can be derived or specified according to empirical evidence and by using several techniques. Cheli and Lemmi (1995), for instance, suggest a "Totally Fuzzy and Relative approach" that matches the membership function to the sample distribution of the quantitative or qualitative variable chosen, so that the membership scores of a given individual will depend totally on his/her relative position in the distribution of a specific indicator in a given context[13]. Alternatively, a suitable membership function can be derived using methods of interpola-

---

[12] These are more properly called "truncated trapezoidal or shouldered fuzzy sets", and are characterized by the fact that left and right "shoulder" (or plateau) regions are associated to the endpoints of the variable (Cox 1994).

[13] More specifically, they define the membership function as $[1-F(y_i)]$, where F is the distribution function of the chosen indicator.

tion of given sample data; or, finally, the least-square method can be applied for fitting data and estimating the membership function parameters[14].

Third, membership grades and functions can be based on the judgement and experience of external experts; for instance, doctors and nutritional experts can be asked to identify nutritional levels and define the membership function completely in terms of a justifiable mathematical formula, or to associate membership degrees for selected levels. Alternatively, and according to the literature on subjective methods of individual well-being assessment, membership scores can be assigned on the basis of the subjective perceptions of the survey respondents. For instance, interviewees could be asked to associate qualitative and quantitative contents to labels such as “low”, “good”, and so forth, ranking them and assigning membership scores within a given range of values (say, from 0 to 1).

What seems to emerge clearly is that whatever criteria one adopts to construct membership functions, they should be able to convey the semantic properties of the underlying concept; the closer the membership function maps the nature and behaviour of our conceptual phenomenon, the better it will reflect the real word that we want to describe.

There is one final aspect of the descriptive phase that I would like to emphasize. As has already been pointed out, the sharp conceptual and technical distinction between poverty and well-being traditionally employed in these sibling fields of analysis is substituted in the capability perspective by a gradual transition between two opposite conditions (e.g., totally in/totally out; fully achieved/fully not achieved), where the whole “in-between” range is understood to be of equal importance.

The assumption underlying such a unifying conceptual framework, of course, significantly affects the standard approach to poverty analysis, calling into question in particular its ex-ante determination of poverty thresholds. Firstly, this framework demands that attention be paid to conditions of overall human well-being, in all their intrinsic, complex, and multidimensional nature, and that they be represented along a continuum, from the situation of the most disadvantaged to that of the most privileged. Secondly, it brings the decision to “draw a line” back to its appropriate realm, i.e. that of value-judgements and social responsibilities, of political decisions and financial constraints.

Focusing attention and public action on specific subgroups of the population characterized by low levels of achievement of well-being clearly requires appropriate technical tools for measuring and monitoring progress and deterioration, but the decisions regarding which subgroups of popula-

---

[14] A classic textbook on fuzzy set theory that includes some basic principles on these methods is, among others, Klir and Yuan (1995).

tion on whom to focus, or what should be considered a "low" level of well-being for a given society, do not seem to be truly "technical" matters.

From this point of view fuzzy methodology would also seem to be the "natural candidate" for representing poverty and well-being as two joint, symmetric aspects of the same problem.

Figures 5.2-5.5 show some alternative specifications that can be used to describe the process of transition between two opposite conditions, identified respectively in terms of "full deprivation" and "full achievement".

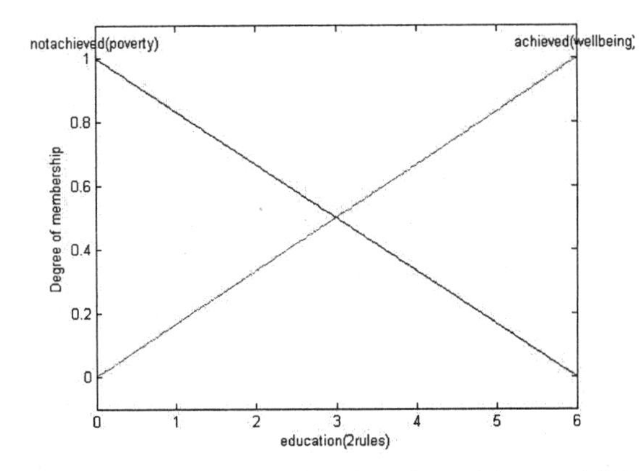

Fig. 5.2. Linear membership function with 2 modalities or rules μ (x; a, b)
a) poverty condition (fully not achieved) = (b-x)/(b-a)
b) well-being condition (fully achieved) = (x-a)/(b-a)
where a and b are parameters associated respectively with the bottom and top rank.

While Figures 5.2 and 5.4 illustrate two symmetric modalities, or rules of specification, determining two fuzzy subsets (not achieved/fully achieved), Figures 5.3 and 5.5 include a third intermediate position, "moderately achieved", thereby increasing the range of specification through the identification of a third fuzzy subset, i.e. the subset of people for whom the functioning of "*being educated*" or "*being nourished*" has been moderately achieved. Shifting these functions and varying the parameters makes it possible to characterize fuzzy sets differently, according to personal and environmental aspects (i.e. elements of the $\mathbf{h}_i$ and $\mathbf{e}_i$ vectors) that can affect the functioning achievement, while maintaining the membership scores in the (0,1) range as a common metric[15].

[15] For instance, the fuzzy set "*being educated*" can be characterized by different membership functions if referred to the young or the elderly. Similarly, nutritional status can be depicted differently for children, adult men or old women.

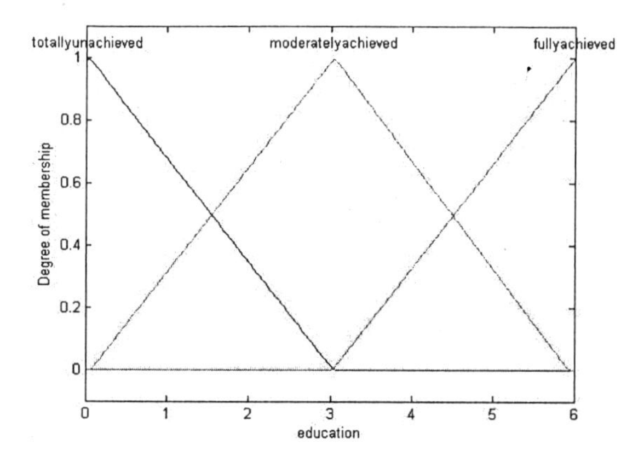

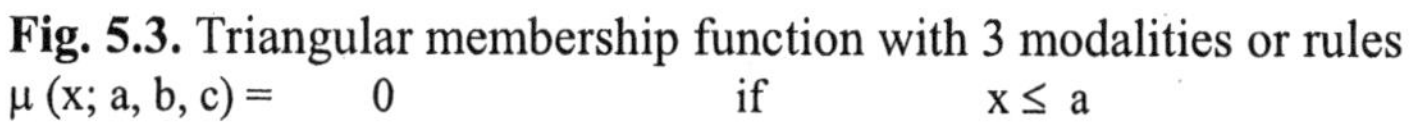

Fig. 5.3. Triangular membership function with 3 modalities or rules

$$\mu\,(x;\,a,\,b,\,c) = \begin{cases} 0 & \text{if} \quad x \leq a \\ (x\text{-}a)/(b\text{-}a) & \text{if} \quad a \leq x \leq b \\ (c\text{-}x)/(c\text{-}b) & \text{if} \quad b \leq x \leq c \\ 0 & \text{if} \quad c \leq x \end{cases}$$

where a and c locate the bottom of the triangle and b locates the "peak".

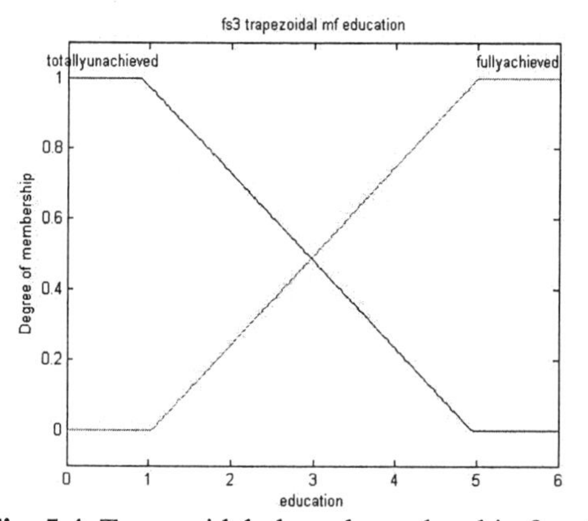

Fig. 5.4. Trapezoidal-shaped membership function with 2 modalities or rules

$$\mu\,(x;\,a,\,b) = \begin{cases} 0 & \text{if} \quad x \leq a \\ (x\text{-}a)/(b\text{-}a) & \text{if} \quad a \leq x \leq b \\ 1 & \text{if} \quad x \geq b \end{cases} \qquad \text{(totally unachieved)}$$

$$\mu\,(x;\,a,\,b) = \begin{cases} 1 & \text{if} \quad x \leq b \\ (b\text{-}x)/(b\text{-}a) & \text{if} \quad b \leq x \leq a \\ 0 & \text{if} \quad x \geq a \end{cases} \qquad \text{(fully achieved)}$$

where a and b are parameters associated respectively with the bottom and top thresholds.

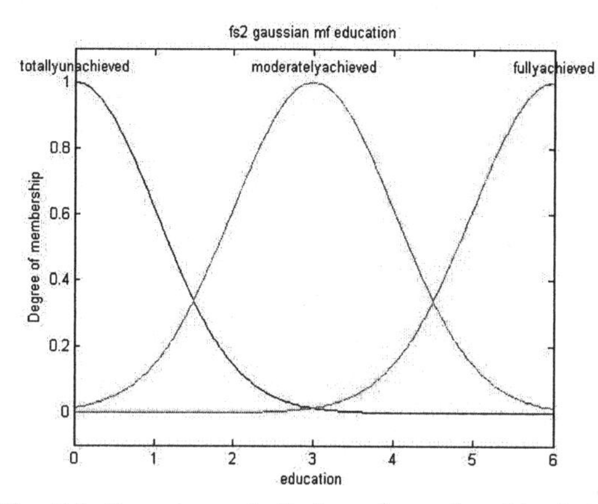

Fig. 5.5. Gaussian or bell-shaped membership function with 3 modalities or rules

μ (x; σ, c)= $\exp\left[-(x-c)^2/2\sigma^2\right]$ where c and σ identify respectively the central value and the standard deviation of the curve.

5.4 Aggregating well-being dimensions through fuzzy operators

Similarly to what happens with conventional or crisp sets, complement, intersection and union operations make it possible to manipulate and combine elementary fuzzy sets. However, since fuzzy sets are not crisply partitioned as are conventional sets, the operators apply on the membership functions, determining membership degrees that, once again, will not be restricted simply to 0 and 1.

I will not discuss here the technical aspects related to these operators, nor their application to the empirical analysis of poverty and well-being [16]. It might be useful, however, to discuss briefly in what sense the wide range of fuzzy set operators available seems to be particularly interesting and meaningful from a capability perspective – how these operators can be applied for aggregating across functionings or well-being dimensions, and how they can be interpreted. I will focus here on complement, intersection and union operators only.

[16] I discussed these aspects in Chiappero-Martinetti (1994, 2000) with reference to the capability approach. There is extensive literature on fuzzy set operators, and basic elements are available in every textbook on fuzzy sets.

5.4.1 Standard complement or negation

If conventional logic complement or negation simply switches membership degrees from 0 to 1 and vice versa, fuzzy logic extends this principle to the whole range of values between 0 and 1. Given fuzzy set A, fuzzy complement ~A is usually defined by the operation $1-\mu_A(x)$ and can be interpreted as the degree to which element x belongs to ~A or, equivalently, does not belong to A.

What it is worth emphasizing from the point of view of the interpretation of the negation operation is that it determines the degree to which an element is complementary to the underlying fuzzy concept, but does not represent its opposite, as is the case in traditional, Boolean logic.

Conceptually speaking, in classical logic a statement must be true or false, black or white, poor or not poor, while fuzzy logic intentionally rejects such bipolar thinking. The fuzzy membership degree to the subset of, for example, "white man", or "Catholic people" is not necessarily or exactly equivalent to the fuzzy membership score to the subset of "non-white man" or "non-Catholic people", given the possibility of belonging, to a certain degree, also to other racial or religious categories.

Formally, in classic logic, the conventional complement identifies crisply defined partitions and satisfies the so-called "law of non-contradiction", according to which the intersection of a set with its complement is an empty set (formally: $A \cap {\sim}A = \varnothing$). Elements must necessarily and precisely be classified in one of the two opposite sets.

The fuzzy complement does not refer to two disjointed domains, but measures the distance between two points in the same domain. A statement can be partially true, many people can be neither black nor white, a functioning can be partially achieved, and so on. The violation of the principle of non-contradiction that occurs for fuzzy sets can be seen in Figures 5.2-5.5, where an overlapping between regions typically exists. An element can belong to two (or more) subsets and not necessarily with the same membership degrees (i.e. the membership functions must not necessarily be symmetric as described in Figures 5.2-5.5).

5.4.2 Standard intersection and standard union

If the intersection of two crisp sets contains the elements that are common to both sets, the corresponding standard intersection operation in fuzzy logic (or logical *and*) is determined by taking the minimum of the two corresponding membership degrees. The standard intersection between two fuzzy sets A and B with membership degrees μ_A and μ_B can thus be indicated as $\mu_{A \cap B} = \min [\mu_A, \mu_B]$.

With reference to our bi-dimensional example in the functionings space, the intersection operator would select the lower values between the two membership degrees related to both fuzzy sets "being educated" and "being nourished", no matter how high the score in the other set might be.

This means that we are implicitly rejecting the hypothesis that a sort of positive compensation or trade-off might be possible between these two dimensions of well-being: both conditions – "being educated" and "being nourished" are necessary conditions for achieving well-being, and a high educational level cannot make up for poor nutritional status. For this reason, the intersection operators can be usefully applied when there is a positive correlation between them, i.e. when the dimensions or attributes are complements (a higher level of education is associated with better achievement in terms of nutritional status).

Conversely, the standard union corresponds to the logical operator *or* in classical logic, and is performed by taking the maximum value between the two membership degrees. Given two fuzzy sets A and B, the standard union can be expressed as $\mu_{A \cap B} = \max [\mu_A, \mu_B]$.

The union operator would choose the higher score related to either fuzzy set – "being educated" or "being nourished" – no matter how low the other score might be. This would correspond to a full compensation of lower degrees of membership by the maximum degree of membership. The higher membership score in the educational or nutritional functioning space would be a sufficient condition for achieving well-being.

Standard union and standard intersection are simple to determine and intuitive; they generalize the classical set theory when the range of values is restricted to 0 and 1 and can be adequate to deal with simple (mainly bi-dimensional) fuzzy models for which a clear correlation between dimensions can be identified. In more complex, n-dimensional spaces, a potential limit of the standard operators (in particular, the standard intersection) is that extreme values can control the overall assessment.

5.4.3 Other common fuzzy sets operators

Other classes of fuzzy set operators, representing equally possible generalization of crisp sets theory, have been suggested (see Zadeh 1965; Giles 1976; Yager 1980; Dubois and Prade 1985). They find their justification in intuitive argumentations as well as empirical and axiomatic explanations. For instance, "weak intersection" (or algebraic product) $\mu_{A \cdot B} = [\mu_A \cdot \mu_B]$ and "weak union" (or algebraic sum) $\mu_{A+B} = [\mu_A + \mu_B - \mu_A \cdot \mu_B]$ admit compensation between A and B and can be adequate when independent well-being dimensions are considered[17].

A bounded difference $\mu_{A\cap B} = \max[0, \mu_A + \mu_B - 1]$ and a bounded sum $\mu_{A\cup B} = \min[1, \mu_A + \mu_B]$ can be applied in the case of a negative correlation between indicators, but they decrease the possibility of "fuzzifying" the extreme values. In fact, the bounded difference acts as a selective filter that admits only high membership values (when the summation of the two membership degrees exceeds 1) while the bounded sum, to the contrary, introduces a relatively low filter with aggregate membership degrees that quickly approach 1.

Finally, in fuzzy set theory logical "and" and logical "or" can be easily generalized by using the averaging operators that explicitly admit the idea of compensation between conflicting (and/or) goals: the resulting trade-offs lie between the minimum and the maximum degree of membership of the aggregated sets. Weighted and unweighted means are examples of averaging operators: while the former makes it possible to assign a different weight to the elementary sets, the latter preserves the symmetry assumption that the aggregated sets are equally important.

The availability of a variety of fuzzy set operators makes it possible to fit the aggregation phase to the conceptual framework and to perform union or intersection operators of various strengths not in a mechanical manner but through a process of continuous adaptation and calibration between techniques and theoretical foundations. The choice of the appropriate set of operators is strongly dependent on the specific context of application, and should be based on empirical verification of different sorts of aggregator operators.

5.5 Assessing multidimensional well-being through fuzzy inference systems

The last phase that remains to be discussed is the passage from the description of single functionings and the aggregation across functionings to an overall assessment of individual well-being, a phase that will be discussed and implemented here making use of fuzzy inference systems (FIS).

---

[17] It should be taken into account, however, that since the membership degree in the aggregated set depends on the number of combined sets, when the algebraic product is used the value of the aggregate membership degree progressively decreases when the number of fuzzy sets increases. This may or may not be a desirable property (Zimmermann 1990).

A fuzzy inference system is essentially a process of inference based on fuzzy production rules (or approximate reasoning) that connects antecedents with consequences in a straightforward manner, making use of membership functions, fuzzy operators, and "if-then" rules.

The architecture of a FIS is based on three main components: a set of inputs, a process of inference, and a final output. Both input and output are crisp, real numbers that go through a fuzzification process.

The set of inputs (x_1, x_2, ..., x_n) constitutes our database and are generally crisp numbers restricted to a specific range of values (amount of calories, ordinal or interval scales, etc.). The first step requires that we fuzzify these inputs determining their membership degrees to each corresponding fuzzy set via membership functions. This step corresponds to the description phase discussed in Sect. 5.3 and requires that we define labels and linguistic attributes for each input variable and determine and calibrate the appropriate membership functions.

The process of inference operates on a set of production rules that connect antecedents ("if") with consequences ("then") using fuzzy operators, thereby producing fuzzy inference[18]. A common rule can have the form:

$$\text{if } (x \text{ is } K) \cdot (y \text{ is } W) \text{ then } z \text{ is } S \tag{5.2}$$

where "·" is a fuzzy connector that takes the form of the AND (intersection) or OR (union) operators. In the case of our functionings achievement this rule could take the form: "if educational level is very low and/or nutritional status is very poor than well-being is very low"[19]. The and/or operators will be chosen from the wide range of operators formulated in fuzzy sets literature and briefly mentioned in Sect. 5.4.

The aggregate output produced by the inference engine is a fuzzy set that, as a final step, must be "defuzzified" so to be expressed by a real number. In our example, the final output that we need is a single number (in the whole range between 0 and 1) that represents the well-being of a

[18] The fuzzy sets literature has formulated many fuzzy inference methods. The FIS applied in this Chapter was suggested by Mamdani and Assilian (1975).

[19] This is evidently a trivial example that generates a 0 value in terms of membership degree to the fuzzy set of "fully achieved well-being" (or, symmetrically, a value equal to 1 to the fuzzy set of "totally not achieved well-being") and corresponds to the conventional, dichotomous case (fully out/fully in). As has already been pointed out, what makes the difference between crisp and fuzzy aggregation operators is that the latter offer a wider range of intersection and union operators and allow "fuzzification" of all intermediate positions between the two extreme conditions.

given individual determined as a synthesis of his or her position in the multidimensional (fuzzy) functionings space[20].

Fig. 5.6 shows some well-being surfaces that can be generated by a fuzzy inference system.

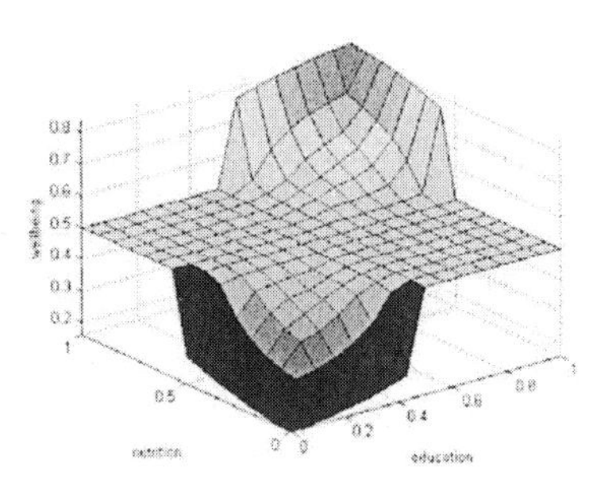

(a) Three triangular mf, standard intersection

(b) Three triangular mf, standard union

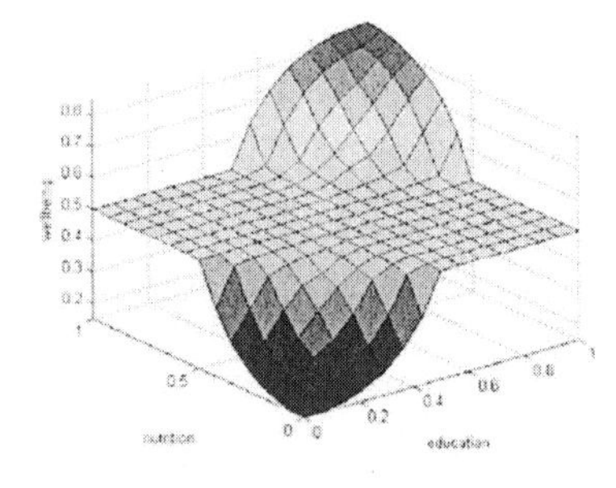

(c) Three Gaussian mf, standard intersection

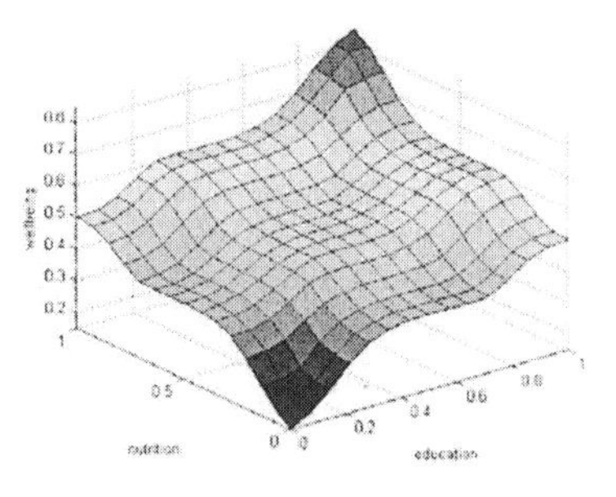

(d) Three Gaussian mf, standard union

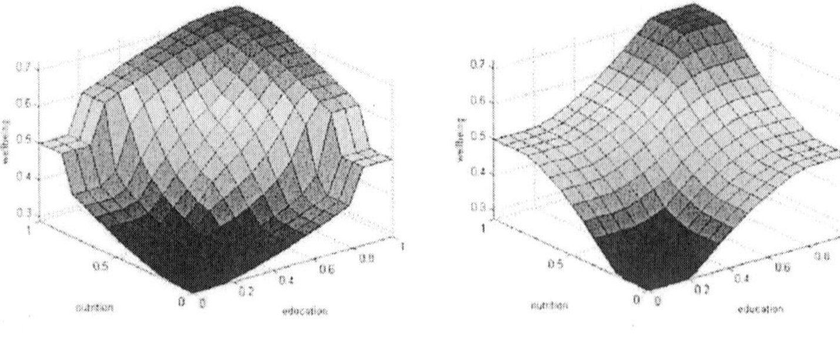

(e) Two trapezoidal mf, standard intersection

(f) Two trapezoidal mf, standard union

[20] Defuzzification methods adopt different criteria for determining the crisp value: the most frequently adopted is the centroid method based on the value that occupies the centre of the area under the curve (Cox 1994; Klir and Folger 1995).

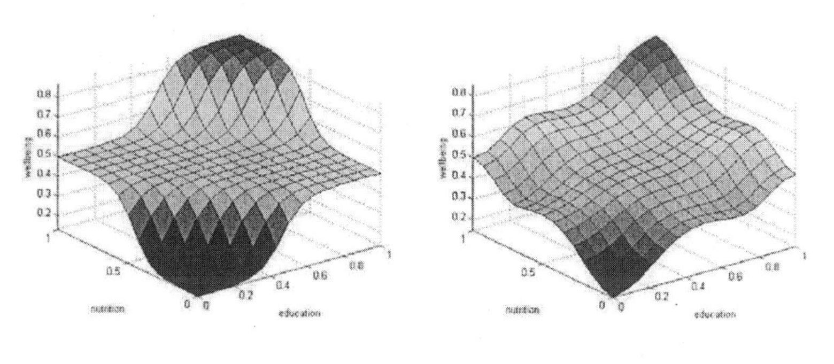

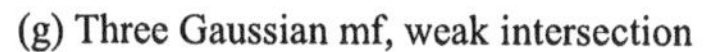

(g) Three Gaussian mf, weak intersection (h) Three Gaussian mf, weak union

Fig. 5.6. Well-being surfaces based on different membership functions and rules of aggregation

The different types of membership functions previously discussed (cfr. triangular, trapezoidal and Gaussian membership functions with two or three modalities presented in Figg. 5.2-5.5) and several common fuzzy aggregators (standard and weak intersection, standard and weak union) have been taken into consideration to determine the overall well-being achievement.

The vertex of these diagrams with coordinates (0,0,0) describes the most disadvantaged condition, while the opposite with coordinates (1,1,1) describes those who are best off. Each point of these surfaces determines a different well-being achievement with respect to the two functionings "being educated" and "being nourished".

Following are some general considerations about the shape of these surfaces. If triangular membership functions determine a sort of intermediate plateau in the well-being surface, the bottom and top thresholds associated with the trapezoidal-shape membership functions generate a sort of linear border in the well-being surface, while the Gaussian functions make the surfaces smoother. A smoother profile is also obtained when a weak intersection operator (based on the product between membership degrees) is applied.

5.6 Conclusion

The aim of this Chapter has been to show in what sense fuzzy set theory can be useful in operationalizing multidimensional well-being analysis based on the capability approach. With their capacity to make formal logic and verbal formulations much closer to one another, and to establish close connections between theory and data analysis, fuzzy methodologies seem to be a useful and flexible tool for operationalizing the capability approach while preserving its richness and complexity.

The Chapter has discussed the three main phases of operationalization – description, aggregation and inference – while seeking to underline the main conceptual and methodological issues that could arise from a capability perspective. What is worth emphasizing is that the application of fuzzy sets methodology does not represent a sort of mechanical exercise or standard algorithm, but requires a fundamental interpretative effort for each step, with the aim of establishing a close link between the contents of the theoretical concepts under examination and their representation through fuzzy set theory.

In order to take full advantage of fuzzy methodologies in the operationalization of the capability approach, many other aspects remain to be investigated – e.g. comparing different methods for constructing membership functions, testing different classes of fuzzy operators, making use of fuzzy inference systems for estimating capabilities instead of achieved functionings, and so forth. However, the affinity between fuzzy methodologies and the capability approach seems to be a very promising one.

References

Alkire S (2002) Valuing freedoms. Sen's capability approach and poverty reduction. Oxford University Press, New York

Baliamoune MN (2004) On the measurement of human well-being: fuzzy set theory and Sen's capability approach. Research paper 2004/16, Wider, Helsinki

Cheli B, Lemmi A (1995) Totally fuzzy and relative approach to the multidimensional analysis of poverty. Economic Notes 24:115-134

Chiappero-Martinetti E (1994) A new approach to evaluation of well-being and poverty by fuzzy set theory. Giornale degli economisti e annali di economia 7-9:367-388

Chiappero-Martinetti E (1996) Standard of living evaluation based on Sen's approach: some methodological suggestion. In: Balestrino A, Carter I (eds) Functionings and capabilities: normative and policy issues. Notizie di Politeia, Milan, pp 37-54

Chiappero-Martinetti E (2000) A multidimensional assessment of well-being based on Sen's functioning approach. Rivista Internazionale di Scienze Sociali 108:207-239

Chiappero-Martinetti E (2005) Complexity and vagueness in the capability approach: strengths or weaknesses? In: Comim F, Qizilbash M, Alkire S (eds) Justice and poverty: essays on operationalizing Amartya Sen's capability approach. Cambridge University Press, Cambridge: forthcoming

Clark D, Qizilbash M (2002) Core poverty and extreme vulnerability in South Africa. The Economics Research Centre, School of Economic and Social Studies, University of East Anglia, Discussion Paper 2002/3

Cox E (1994) The fuzzy systems handbook. Academic Press, Chestelnut Hill, Ma

Dubois D, Prade H (1985) A review of fuzzy set aggregation connectives. Information Sciences 36:85-121

Giles R (1976) Lukasiewicz logic and fuzzy theory. Intern J Man-Mach Stud 8:313-327

Klir GJ, Yuan B (1995) Fuzzy sets and fuzzy logic: theory and applications. Prentice Hall, New Jersey

Kuklys W (2005) Amartya Sen's Capability Approach. Theoretical Insights and Empirical Applications. Springer, Berlin

Lelli S (2001) Factor Analysis vs. Fuzzy Sets Theory: Assessing the Influence of Different Techniques on Sen's Functioning Approach. Discussion Paper Series DPS 01.21, November 2001, Center for Economic Studies, Catholic University of Louvain

Mamdani EH, Assilian S (1975) An experiment in linguistic synthesis with a fuzzy logic controller. International Journal Man-Machine Studies 7:1-13

Nussbaum M (2003) Capabilities as fundamental entitlements: Sen and Social Justice. Feminist Economics 9:33-59

Ragin CC (2000) Fuzzy set social science. The University of Chicago Press, Chicago

Robeyns I (2003) Sen's capability approach and gender inequality: selecting relevant capabilities. Feminist Economics 9:61-92

Robeyns I (2005) The capability approach: a theoretical survey. Journal of Human Development 6:93-114

Sen AK (1985) Commodities and capabilities. North Holland, Amsterdam

Sen AK (1992) Inequality reexamined. Clarendon Press, Oxford

Sen AK (2004) Capabilities, lists, and public reason: continuing the conversation. Feminist Economics 10:77-80

Yager RR (1980) On a general class of fuzzy connectives. Fuzzy Set and Systems, 4:235-242

Zadeh LA (1965) Fuzzy sets. Information and Control 8:338-353

Zimmermann HJ (1990) Fuzzy set theory and its applications. Kluwer Academic Publishers, Boston

6 Multidimensional and Longitudinal Poverty: an Integrated Fuzzy Approach

Gianni Betti, Bruno Cheli, Achille Lemmi, Vijay Verma

Gianni Betti, Achille Lemmi, Vijay Verma
Dipartimento di Metodi Quantitativi, University of Siena
Bruno Cheli
Dipartimento di Statistica e Matematica applicata all'Economia, University of Pisa

6.1 Introduction

This Chapter is a contribution to the analysis of deprivation seen as a multi-dimensional condition, and in the longitudinal context. Multi-dimensionality involves both monetary and diverse non-monetary aspects – the former as the situation, either absolute or relative to the average standard, of low income, and the latter as a lack of access to other resources, facilities, social interactions and even individual attributes determining life-style. Persistence and movement over time is an equally important aspect of the intensity of deprivation, requiring longitudinal study at the micro level and in the aggregate.

Most of the methods designed for the analysis of poverty share two limitations: i) they are unidimensional, i.e. they refer to only one proxy of poverty, namely low income or consumption expenditure; ii) they need to dichotomize the population into the *poor* and the *non-poor* by means of the so called *poverty line*.

Nowadays many authors recognize that poverty is a complex phenomenon that cannot be reduced solely to monetary dimension. This leads to the need for a multidimensional approach that consists in extending the analysis to a variety of non-monetary indicators of living conditions. If multidimensional analyses are increasingly feasible as the available information increases, it was the development of multidimensional approaches that in turn stimulated the surveying of a variety of aspects of living conditions.

By contrast, however, little attention has been devoted to the second limitation of the traditional approach, i.e. the rigid poor/non-poor dichotomy, with the consequence that most of the literature on poverty measurement continues to be based on the use of poverty thresholds. Yet it is undisputable that such a clear-cut division causes a loss of information and removes the nuances that exist between the two extremes of substantial welfare on the one hand and distinct material hardship on the other. In other words, poverty should be considered as a *matter of degree* rather than as an attribute that is simply present or absent for individuals in the population.

An early attempt to incorporate this concept at methodological level (and in a multidimensional framework) was made by Cerioli and Zani (1990) who drew inspiration from the theory of Fuzzy Sets initiated by Zadeh (1965).

Given a set X of elements $x \in X$, any fuzzy subset A of X is defined as follows: $A = \{x, \mu_A(x)\}$, where $\mu_A(x)$: $X \rightarrow [0,1]$ is called the *membership function* (*m.f.*) in the fuzzy subset A. The value $\mu_A(x)$ indicates the degree of membership of *x* in *A*. Thus $\mu_A(x) = 0$ means that x does not belong to *A*, whereas $\mu_A(x) = 1$ means that *x* belongs to *A* completely. When $0 < \mu_A(x) < 1$, *x* partially belongs to *A* and its degree of membership in *A* increases in proportion to the proximity of $\mu_A(x)$ to 1.

Cerioli and Zani's original proposal was later developed by Cheli and Lemmi (1995) giving origin to the so called *Totally Fuzzy and Relative* (TFR) approach. Both methods have been applied by a number of authors subsequently, with a preference for the TFR version[1], and in parallel the same TFR method was refined by Cheli (1995) who also used it to analyze poverty in fuzzy terms in the dynamic context represented by two consecutive panel waves.

From this point on, the methodological implementation of this approach has developed in two directions, with somewhat different emphasis despite their common orientation and framework. The first of these is typified by the contributions of Cheli and Betti (1999) and Betti, Cheli and Cambini (2004), focusing more on the time dimension, in particular utilizing the tool of transition matrices. The second, with the contributions of Betti and Verma (1999, 2002, 2004) and Verma and Betti (2002), has focused more on capturing the multi-dimensional aspects, developing the concepts of "manifest" and "latent" deprivation to reflect the intersection and union of different dimensions.

In this Chapter we draw on the state-of-the-art of these developments, to integrate them in the form of, what may be called, an "Integrated Fuzzy and Relative" (IFR) approach to the analysis of poverty and deprivation.

[1] For instance, Chiappero Martinetti (2000), Clark and Quizilbash (2002) and Lelli (2001) use the TFR method in order to analyze poverty or well-being according to Sen's capability approach.

The concern of the chapter is primarily methodological. We re-examine two important aspects introduced by the use of fuzzy measures, namely:

(i) the choice of membership functions i.e. quantitative specification of individuals' or households' degrees of poverty and deprivation, given the level and distribution of income and other aspects of living conditions of the population; and

(ii) the choice of rules for the manipulation of the resulting fuzzy sets, rules defining their complements, intersection, union and aggregation.

In relation to (i), we note the relationship of the proposed fuzzy monetary measure with the Lorenz curve and the Gini coefficient. Certain conceptual and theoretical aspects concerning fuzzy set logic and operations pertinent for the definition of multidimensional measures of deprivation are then clarified, and utilized in the construction of a number of such measures.

The need for (ii) arises because, for longitudinal analysis of poverty using the fuzzy set approach, we need *joint membership functions* covering more than one time period, which have to be constructed on the basis of the series of cross-sectional membership functions over those time periods. We propose a general rule for the construction of fuzzy set intersections, that is, a rule for the construction of longitudinal poverty measures from a sequence of cross-sectional measures. On the basis of the results obtained, various fuzzy poverty measures over time can be constructed as consistent generalizations of the corresponding conventional (dichotomous) measures. Examples are rates of any-time, persistent and continuous poverty, distribution of persons and poverty spells according to duration, rates of exit and re-entry into the state of poverty, etc.

6.2 Income poverty

Diverse "conventional" measures of monetary poverty and inequality are well-known and are not discussed here. Here we will focus on only the most commonly used measure, namely the proportion of a population classified as "poor" in purely relative terms on the following lines. To dichotomize the population into the "poor" and the "non-poor" groups, each person i is assigned the equivalised income y_i of the person's household. Persons with equivalised income below a certain threshold or poverty line (such as 60% of the median equivalised income as adopted by Eurostat) are considered to be poor, and the others as non-poor. The conventional income poverty rate (the head count ratio, H) is the proportion of the population below the poverty line.

Apart from the various methodological choices involved in the construction of conventional poverty measures, the introduction of fuzzy measures brings in *additional* factors on which choices have to be made. As noted, these concern at least two aspects: the choice of *membership functions*; and the choice of *rules for manipulation* of the resulting fuzzy sets. To be meaningful both these choices must meet some basic logical and substantive requirements. It is also desirable that they be useful in the sense of elucidating aspects of the situation not captured (or not captured as adequately) by the conventional approach.

We begin with the issue of choice of the poverty membership function (m.f.). In the conventional head count ratio H, the m.f. may be seen as $\mu(y_i)=1$ *if* $y_i<z$, $\mu(y_i)=0$ *if* $y_i \geq z$, where y_i is equivalised income of individual i, and z is the poverty line. In order to move away from the poor/non-poor dichotomy, Cerioli and Zani (1990) proposed the introduction of a transition zone (z_1-z_2) between the two states, a zone over which the m.f. declines from 1 to 0 linearly.

In the TFR approach, Cheli and Lemmi (1995) define the m.f. as the distribution function of income, normalized (linearly transformed) so as to equal 1 for the poorest and 0 for the richest person in the population. The mean of m.f. defined in this way is always 0.5, by definition. It is desirable, however, to make this mean represent the average level of poverty or deprivation in the population, just as H in the conventional approach.

In order to make this mean equal to some specified value (such as 0.1) so as to facilitate comparison with the conventional poverty rate, Cheli (1995) takes the m.f. as normalized distribution function, raised to some power $\alpha \geq 1$:

$$\mu_i = (1-F_i)^{\alpha} = \left(\sum_{\gamma=i+1}^{n} w_{\gamma} \Big/ \sum_{\gamma=2}^{n} w_{\gamma} \right)^{\alpha}; \quad \mu_n = 0 \tag{6.1}$$

where F_i is the income distribution function and w_γ is the sample weight of individual of rank γ (1 to n) in the ascending income distribution.

Increasing the value of this exponent implies giving more weight to the poorer end of the income distribution: empirically, large values of the m.f. would then be concentrated at that end, making the propensity to income poverty sensitive to the *location* of the poorer persons in the income distribution. Beyond that, the choice of the value of α is essentially arbitrary, or at best based on some external consideration: this is unavoidable since any method for the quantification of the extent of poverty is inevitably based on the arbitrary choice of some parameter (Hagenaars 1986). Later Cheli and Betti (1999) and Betti and Verma (1999) have chosen the parameter α so that the mean of the m.f. is equal to the head count ratio computed for the official poverty line. In this way we avoid explicit choice of α, by adapting to the political choice which is implicit in the poverty line. Moreover, comparison between the conventional and fuzzy measures is facilitated. Betti and Verma (1999) have used a somewhat refined version of the above formulation (6.1) in order to define what they called the Fuzzy Monetary indicator (FM):

$$\mu_i = FM_i = (1-L_i)^\alpha = \left(\sum_{\gamma=i+1}^{n} w_\gamma y_\gamma \Big/ \sum_{\gamma=2}^{n} w_\gamma y_\gamma \right)^\alpha ; \quad \mu_n = 0 \tag{6.2}$$

where y_γ is the equivalised income and L_i represents the value of the Lorenz curve of income for individual i. In other terms, $(1-L_i)$ represents the *share of the total equivalised income* received by all individuals who are less poor than the person concerned. It varies from 1 for the poorest, to 0 for the richest individual. $(1-L_i)$ can be expected to be a more sensitive indicator of the actual disparities in income, compared to $(1-F_i)$ which is simply the *proportion of individuals* less poor than the person concerned. It may be noted that while the mean of $(1-F_i)$ values is ½ by definition, the mean of $(1-L_i)$ values equals $(1+G)/2$, where G is the Gini coefficient of the distribution.

Here we propose a new measure which combines the TFR approach of Cheli and Lemmi (1995) and the approach of Betti and Verma (1999) into an "Integrated Fuzzy and Relative" (IFR) approach, which takes into account both the *proportion of individuals* less poor than the person concerned, and the *share of the total equivalised income* received by all individuals less poor than the person concerned. We define this measure as:

$$\mu_i = FM_i = (1-F_i)^{\alpha-1}\left[1-L_i\right] =$$

$$= \left(\sum_{\gamma=i+1}^{n} w_\gamma \Big/ \sum_{\gamma=2}^{n} w_\gamma \right)^{\alpha-1} \left(\sum_{\gamma=i+1}^{n} w_\gamma y_\gamma \Big/ \sum_{\gamma=2}^{n} w_\gamma y_\gamma \right); \quad \mu_n = 0 \tag{6.3}$$

where, again, parameter α may be chosen so that the mean of these measures, FM, equals the head count ratio H:

$$FM = \frac{\alpha + G_\alpha}{\alpha(\alpha+1)} = H \tag{6.4}$$

It is important to note that the Fuzzy Monetary (FM) measure as defined above is expressible in terms of the generalized Gini measures G_α, which is a generalization of the standard Gini coefficient (for $\alpha=1$). In the continuous case it is defined as:

$$G_\alpha = \alpha(\alpha+1)\int_0^1 (1-F)^{\alpha-1}(F-L)\,dF . \tag{6.5}$$

This measure weights the distance $(F - L)$ between the line of perfect equality and the Lorenz curve by a function of the individual's position in the income distribution, giving more weight to its poorer end.

6.3 Non-monetary deprivation ("Fuzzy Supplementary")

In addition to the level of monetary income, the standard of living of households and persons can be described by a host of indicators. Quantification of and putting together diverse indicators of deprivation involves a number of steps, models and assumptions. Specifically, decisions are required with regard to assigning numerical values to the ordered categories, weighting the score to construct composite indicators, choosing their appropriate distributional form, and scaling the resulting measures in a meaningful way.

Choice and grouping of indicators

Firstly, from the large set which may be available, a selection has to be made of indicators which are substantively meaningful and useful for the analysis of deprivation. This is a substantive as well as a statistical question. Secondly, it is useful to identify the underlying dimensions and to group the indicators accordingly. Taking into account the manner in which different indicators cluster together adds to the richness of the analysis; ignoring such dimensionality can result in misleading conclusions (Whelan et al. 2001).

Assigning numerical values to ordered categories

Individual items indicating non-monetary deprivation often take the form of simple "yes/no" dichotomies (such as the presence or absence of enforced lack of certain goods or facilities), or sometimes ordered polytomies. Perhaps the simplest scheme for assigning numerical values to categories is by assuming that the ranking of the categories represents an equally-spaced metric variable (Cerioli and Zani 1990). An alternative which has been proposed is replacing the simple ranking of the categories with their distribution function in the population (Cheli and Lemmi 1995).

Weighting for constructing composite measures

When aggregating several indicators at macro level, an early attempt to choose an appropriate weighting system was made by Ram (1982), using principal component analysis, which was also adopted by Maasoumi and Nickelsburg (1988). For the construction of fuzzy measures, however, it is necessary to weight and aggregate items at the *micro level*. Nolan and Whelan (1996) adopted factor analysis for this purpose. In order also to give more weight to more widespread items, Cerioli and Zani (1990) specified the weight of any item as a function of the proportion deprived of the item. Another very important principle that the weighting system should satisfy is that of avoiding redundancy, that is, limiting the influence of those indicators that are highly correlated. To this effect, Betti and Verma (1999) proposed the item weights to comprise two factors. The first factor is determined by the variable's power to differentiate among individuals in the population, that is, by its dispersion: this may be taken as proportional to the coefficient of variation of deprivation score for the variable concerned. The second factor is taken as a function of the correlation of any item with other items, in such a manner that it is not affected by the introduction of variables entirely uncorrelated with the item concerned, but is reduced proportionately to the number of highly correlated variables present.

Functional form of the distribution

Of course, the numerical values for composite indicators of deprivation as obtained above may be directly used as fuzzy degrees of membership, as has been done by a number of authors. Betti and Verma (1999) proposed instead to treat the non-monetary scores in a way entirely analogous to that for monetary poverty measures, described in the previous section. On the basis of this approach, the function corresponding to equation (6.2) would be:

$$\mu_i = FS_i = \left(1-F_{(S),i}\right)^{\alpha_S - 1}\left(1-L_{(S),i}\right);\ \alpha_S \geq 1, \qquad (6.6)$$

where $F_{(S),i}$ represents the distribution function of the overall supplementary deprivation (S) evaluated for individual i, and $L_{(S),i}$ the value of the Lorenz curve of S for individual i, and α_S is a parameter to be determined[2].

[2] The above approach to combining diverse indicators of non-monetary deprivation treats them as additive. The same methodology can be applied for constructing separate measures for different dimensions of non-monetary deprivation, such as those concerning life-style, housing or the environment (Eurostat 2002; Nolan and Whelan 1996). In either case, alternative forms of aggregation are also possible, such as adding the scores separately within dimensions of deprivation, and then aggregating the dimension-specific scores using some other methodology such as fuzzy intersections and unions.

Scaling of the measures

Strictly, the scale of the deprivation measures so constructed remains arbitrary. From a substantive point of view, Betti and Verma (1999) propose to determine α_s so as to make the overall non-monetary deprivation rate numerically identical to the *monetary poverty* rate H.

This completes the specification of the fuzzy m.f. of deprivation.

6.4 Fuzzy set operations appropriate for the analysis of poverty and deprivation

6.4.1 Multidimensional measures

In the previous sections we have considered poverty as a fuzzy state and defined measures of its degree in different dimensions, namely: monetary, overall non-monetary, and possibly concerning particular aspects of life. In multidimensional analysis it is of interest to know the extent to which deprivation in different dimensions tends to overlap for individuals. Similarly, in longitudinal analysis it would be of interest to know the extent to which the state of poverty or deprivation persists over time for the person concerned. Such analyses require the specification of rules for the manipulation of fuzzy sets.

As a concrete example, consider deprivation in two dimensions: monetary poverty and supplementary (overall non-monetary) deprivation that we denote by *m* and *s* respectively, each of them being characterized by two opposite states - labeled as 0 (non-deprived) and 1 (deprived) - that correspond to a pair of sets forming a fuzzy partition. Any individual i belongs *to a certain degree* to each of the four sets: the two cross-sectional sets m and s, and their complements. Since fuzzy sets 0 and 1 are complementary, having defined the degree of membership in one as FM_i or FS_i, it is straightforward (and necessary) to calculate the membership in its complementary set as $(1 - FM_i)$ or $(1 - FS_i)$, respectively.

In the conventional approach, a joint analysis of monetary and non-monetary deprivation (both seen as dichotomous characteristics) is carried out by assigning each individual to one (and only one) of the four sets representing the intersections $m \cap s$ ($m = 0,1$; $s = 0,1$). This can be viewed as individual membership functions in the four sets such that, for a given individual, the membership equals 1 in one of the sets and equals 0 in the three remaining sets. For any particular set, the mean value of the individual membership functions is simply the proportion of individuals in the category corresponding to that set.

Viewed in this way, these "degrees of membership" in the four cross-sectional sets sum to 1 for any individual. In a similar way, we view these fuzzy memberships of an individual to form "fuzzy partitions", which must sum to 1 over the four sets[3]. More precisely, denoting by μ_{ims} the degree of membership in $m \cap s$ (m $\in$ [0,1] ; s $\in$ [0,1]) of individual i, the marginal constraints specified in Table 6.1 must be satisfied. The quantity μ_{ims} represents a measure of the extent to which the individual is affected by the particular combination of states (*m*,*s*).

Table 6.1. Situation of a generic individual i seen in fuzzy terms: membership functions for the four intersection sets and for the marginals

		Non-monetary deprivation (s)		
	poverty status	non-poor (0)	poor (1)	total
Monetary deprivation (m)	non-poor (0)	μ_{i00}	μ_{i01}	$1 - FM_i$
	poor (1)	μ_{i10}	μ_{i11}	FM_i
	total	$1 - FS_i$	FS_i	1

6.4.2 Definition of poverty measures according to both monetary and non-monetary dimensions

Our main goal is to find a specification of μ_{ims} that is *the most appropriate to our purpose of analyzing poverty and deprivation.* In this respect, a most important consideration is the following.

Fuzzy set operations are a generalization of the corresponding crisp set operations in the sense that the former reduce to (exactly reproduce) the latter when the fuzzy membership functions, being in the whole range [0,1], are reduced to a {0,1} dichotomy. There is, however, *more than one way in which the fuzzy set operations can be formulated*, each representing an equally valid generalization of the corresponding crisp set operations. The choice among alternative formulations has to be made primarily on substantive grounds: some options are more appropriate (meaningful, convenient) than others, depending on the context and objectives of the appli-

[3] If for each unit in the population, its membership μ_i in a certain set is decomposed into components μ_{ik} such that $\mu_i = \Sigma_k \mu_{ik}$, then the μ_{ik} values constitute m.f.'s corresponding to fuzzy partitions of the original set.

cation. While the rules of fuzzy set operations cannot be discussed in this Chapter in any detail, it is essential to clarify their application *specifically for the study of poverty and deprivation.*

Since fuzzy sets are completely specified by their membership functions, any operation with them (such as union, intersection, complement or aggregation) is defined in terms of the membership functions of the original fuzzy sets involved. As an example, membership μ_{i11} of Table 6.1 is a function of FM_i and FS_i and might be more precisely written as μ_{i11} (FM_i, FS_i). However in the following discussion it will be convenient to use the following simplified notation: (a,b) for the membership functions of two sets for individual i (subscript i can be dropped when not essential), where $a=FM_i$ and $b=FS_i$; also $s_1=\min(a,b)$ and $s_2=\max(a,b)$. We also denote by $\bar{a}=1-a$, $a\cap b$ and $a\cup b$ the basic set operations of complementation, intersection and union, respectively[4]. Table 6.2 shows three commonly-used groups of rules – termed *Standard*, *Algebraic* and *Bounded* (Klir and Yuan, 1995) – specifying fuzzy intersection and union. Such rules are "permissible" in the sense that they satisfy certain essential requirements such as reducing to the corresponding crisp set operations with dichotomous variables, satisfying the required boundary conditions, being monotonic and commutative, etc.

For our application, a most important observation is that the Standard fuzzy operations provide the *largest* (the most loose, the weakest) intersection and by contrast the *smallest* (the most tight or the strongest) union among all the permitted forms. It is for this reason that they have been labeled as i_{max} and u_{min} in Table 6.2. *It is this factor which makes it inappropriate to use the Standard set operations uniformly throughout in our application to poverty analysis.* In fact, if the Standard operation were applied to all the four intersections of Table 6.1, their sum would exceed 1 and the marginal constraints would not be satisfied[5].

Now it can easily be verified that the Algebraic form, applied to all the four intersections, is the *only one* which satisfies the marginal constraints. But despite this numerical consistency, we *do not* regard the Algebraic form to give results which, for our particular application, would be generally acceptable on intuitive or substantive grounds. In fact, if we take the liberty of viewing the fuzzy propensities as probabilities, then the Algebraic product rule $a\cap b$ as the joint probability, $a\cap b=a*b$ implies zero correlation between the two forms of deprivation, which is clearly at

[4] This is a short-hand notation for the following. If, for example, a and b refer to an individual's degrees of memberships in sets A and B respectively, then we write the person's degree of membership of set $A\cap B$ as $a\cap b$.

[5] For details, see Betti and Verma (2004).

variance with the high positive correlation we expect in the real situation for *similar* states. The rule therefore seems to provide an unrealistically *low* estimate for the resulting membership function for the intersection of two similar states. The Standard rules, giving higher overlaps (intersections) are more realistic for (a,b) representing similar states.

Table 6.2. Some basic forms of fuzzy operations

	Intersection	Union
Type of operation	$a \cap b$	$a \cup b$
S (standard)	$\min(a,b)=i_{max}$	$\max(a,b)=u_{min}$
A (algebraic)	a*b	a+b-a*b
B (bounded)	max(0,a+b-1)	min(1,a+b)

By contrast, in relation to *dissimilar* states $(\bar{a}, b)$ and $(a, \bar{b})$ (lack of correspondence between deprivations in two dimensions), it appears that the Algebraic rule (and hence also the Standard rule) tends to give *unrealistically high* estimates for the resulting membership function for the intersection. The reasoning similar to the above applies: in real situations, we expect large negative correlations (hence reduced intersections) between *dissimilar* states in the two dimensions of deprivation. In fact, it can be seen by considering some particular numerical values for $(\bar{a}, b)$ or $(a, \bar{b})$ that Bounded rule, for instance, gives much more realistic results for dissimilar states.

Given the preceding considerations, the specification of the fuzzy intersection $a \cap b$ that appears to be the most reasonable for our particular application and that satisfies the above mentioned marginal constraints is of a "Composite" type as follows (Betti and Verma 2004):

- For sets representing *similar states* - such as the presence (or absence) of both types of deprivation - the Standard operations (which provide larger intersections than Algebraic operations) are used.
- For sets representing *dissimilar states* - such as the presence of one type but the absence of the other type of deprivation - we use the Bounded operations (which provide smaller intersections than Algebraic operations).

By applying this composite intersection the marginal constraints of Table 6.1 are specified as shown in Table 6.3. Note that with this operation the propensity to the deprived in <u>at least one</u> of the two dimensions equals

$\max(FM_i, FS_i)$, which can be viewed as any of the three entirely equivalent forms:

- as the complement of cell "0-0" in Table 6.3, or
- as the sum of the membership functions in the other three cells, or
- as the union of (FM_i, FS_i) under the Standard fuzzy set operations.

Table 6.3. Joint measures of deprivation according to the Betti-Verma Composite operation

		Non-monetary deprivation (s)		
	poverty status	non-poor (0)	poor (1)	total
Monetary deprivation (m)	non-poor (0)	$\min(1-FM_i, 1-FS_i)$ $=1-\max(FM_i, FS_i)$	$\max(0, FS_i-FM_i)$	$1-FM_i$
	poor (1)	$\max(0, FM_i-FS_i)$	$\min(FM_i, FS_i)$	FM_i
	total	$1-FS_i$	FS_i	1

Figure 6.1 illustrates the Composite set operations graphically. *Such a representation is fundamental to the development and illustration of the methodology presented in this Chapter.* In the figure, the degree of membership in the "universal set" X is represented by a rectangle of unit length, and the individual's memberships on the two subsets (say, $0 \le a \le 1$, $0 \le b \le 1$, and their complements) have been placed within it. Different forms of fuzzy set operations (Table 6.2) are reproduced by different placements of the subset memberships within the rectangle for X. The figure shows intersections; fuzzy set unions can be similarly represented. The Standard form, appropriate for *similar* sets, is represented by placing the two memberships (a,b) on the *same base*, so that their intersection is min(a,b), and union is max(a,b). In the Bounded form, appropriate for *dissimilar* sets, the two sets are placed at the *opposite ends* of X, thus their intersection is max(0, a+b-1) and union is min(1, a+b), exactly as required from Table 6.2. It can be seen that the Algebraic form is represented by placing membership (b) symmetrically over memberships (a) and (non-a), i.e. each of the two receiving a proportionate share of (b), respectively a*b and (1-a)*b. Hence a*b is the intersection, while the union is (a+b-a*b). Generally, by moving one set membership higher than the other within X, the overlap (intersection) is reduced, and the underlay (union) increased.

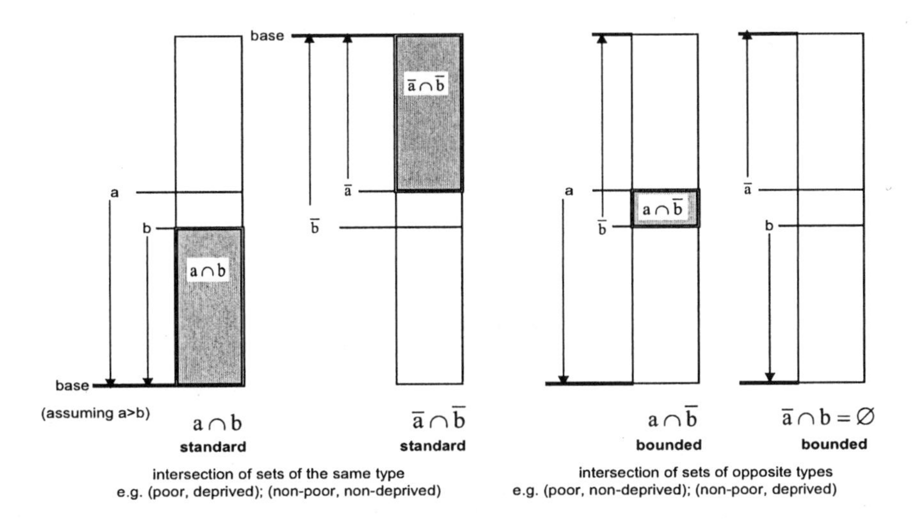

Fig. 6.1. The Composite fuzzy set operations: a graphical representation of intersections

6.4.3 Income poverty and non-monetary deprivation in combination: Manifest and Latent deprivation

The two measures – FM_i the propensity to income poverty, and FS_i the overall non-monetary deprivation propensity – may be combined to construct composite measures which indicate the extent to which the two aspects of income poverty and non-monetary deprivation overlap for the individual concerned. These measures are as follows.

M_i Manifest deprivation,

representing the propensity to both income poverty and non-monetary deprivation simultaneously.

L_i Latent deprivation,

representing the individual being subject to at least one of the two, income poverty and/or non-monetary deprivation.

Once the propensities to income poverty (FM_i) and non-monetary deprivation (FS_i) have been defined at the individual level (i), the corresponding combined measures are obtained in a straightforward way, using the Composite set operations. These individual propensities can then be averaged to produce the relevant rates for the population. The Manifest deprivation propensity of individual i is the intersection (the smaller) of the two (similar) measures FM_i and FS_i:

$$M_i = \min(FM_i, FS_i). \qquad (6.7)$$

Similarly, the Latent deprivation propensity of individual i is the complement of the intersection indicating the absence of both types of deprivation:

$$L_i = 1 - \min(\overline{FM}_i, \overline{FS}_i) = \max(FM_i, FS_i) \qquad (6.8)$$

which turns out to be simply the union (the larger) of the two measures FM_i and FS_i under the Standard operation.

From empirical experience (Betti and Verma 2002; Betti et al. 2005a), it appears that the degree of overlap between income poverty and non-monetary deprivation at the level of individual persons tends to be *higher* in *poorer* areas, and lower in richer areas. A useful indicator in this context is the Manifest deprivation index defined as a percentage of Latent deprivation index; in theory, this ratio varies from 0 to 1. When there is no overlap (i.e., when the subpopulation subject to income poverty is entirely different from the subpopulation subject to non-monetary deprivation), Manifest deprivation rate and hence the above mentioned ratio equals 0. When there is complete overlap, i.e., when each individual is subject to exactly the same degree of income poverty and of non-monetary deprivation ($FM_i = FS_i$), the Manifest and Latent deprivation rates are the same and hence the above mentioned ratio equals 1.

6.5 On longitudinal analysis of poverty conceptualized as a fuzzy state

6.5.1 Longitudinal application of the Composite fuzzy operation

The procedure developed above to represent multi-dimensional aspects of deprivation extends directly to the representation of its longitudinal aspects: in mathematical terms the two are in fact identical. This can be seen from Table 6.4 which shows persistence and transitions in the state of poverty over two time periods.

In place of the two dimensions of deprivation (monetary and non-monetary), here we have fuzzy sets representing the state of poverty at two times. Persistent poverty (row 2 of Table 6.4), for instance, corresponds to "manifest" deprivation defined in the previous section, and "ever in poverty" (row 5) to "latent" deprivation. Similarly, the propensity to exit from poverty (row 3) is given by the intersection of sets representing two dissimilar states, namely set "poor" at time 1 and set "non-poor" at time 2; to these, the Bounded operations apply.

Table 6.4. Longitudinal measures of interest over two time periods for individual i

	Measure	Membership function	Description
1	Never in poverty	$\bar{a}_i \cap \bar{b}_i = 1 - \max(a_i, b_i)$	Poverty at *neither* of the two years
2	Persistent in poverty	$a_i \cap b_i = \min(a_i, b_i)$	Poverty at *both* of the years
3	Exiting from poverty	$a_i \cap \bar{b}_i = \max(0, a_i - b_i)$	Poverty at time 1, but non-poverty at time 2
4	Entering into poverty	$\bar{a}_i \cap b_i = \max(0, b_i - a_i)$	Non-poverty at time 1, but poverty at time 2
5	Ever in poverty	$a_i \cup b_i = \max(a_i, b_i)$	Poverty at *at least one* of the two years

6.5.2 The general procedure

We need procedures which can handle in a consistent and realistic manner the analysis of poverty at any number of time periods (and also for any number of dimensions of deprivation).

Let, for a series of cross-sections (1,...t,...T), each person's propensity to be in poverty (i.e. the person's membership in the set "poor") be given as $(\mu_1, \mu_2, \ldots, \mu_T)$, $\mu_t \in [0,1]$. We also define the complements of the above at each time, i.e. the membership function (m.f.) in the set "non-poor" as $\bar{\mu}_t = 1 - \mu_t$. The above cross-sectional measures generate 2^T longitudinal sequences of length T, in which any element t can take one of two values, μ_t and its complement $\bar{\mu}_t = (1 - \mu_t)$.

Figure 6.2 provides an example of one such sequence. An individual's propensities to poverty (and their complements, propensities to non-poverty) over 6 time periods are represented. Given these cross-sectional propensities (degrees of membership), we need rules to specify the joint membership function (j.m.f.) for any specified longitudinal sequence of particular states, for example of sets "poor" at times (1, 4 and 5), and of sets "non-poor" at the remaining times (2, 3 and 6). These sets of interest are represented by shaded rectangles in Figure 6.2. Note that, as in the case of Figure 6.1, sets representing the same state (e.g., "poverty") are placed on the same base, and those representing the opposite state (e.g., "non-poverty") are placed at the opposite end.

The figure immediately gives the required intersection, i.e. the individual's joint membership for the particular longitudinal sequence: it is simply the overlap (if any) between the smallest of the memberships in the "poor" set, and the largest of the memberships in the "non-poor" set. Clearly, the time-ordering of the various cross-sections is entirely irrelevant in this conceptualization. The result can be seen more clearly by ordering the cross-sections according to the size of the memberships, as shown at the right in the figure.

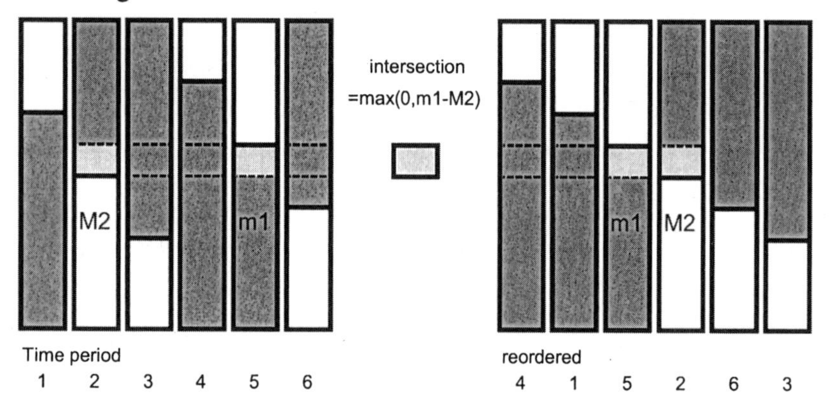

Fig. 6.2. Example of degrees of membership for a longitudinal sequence of "poor" and "non-poor" sets

Returning to the general case, let S(1,2,….,T) be a *particular pattern* of T "poor" and "non-poor" sets for which the j.m.f. is required. Let the elements (cross-sectional sets) of this pattern be grouped into two parts: $S_1 = (....,t_1,....)$, $S_2 = (....,t_2,....)$, where t_1 indicates any of T_1 elements of the same type (say, "poor") in the first group, and t_2 any of T_2 elements of the group of the opposite type ("non-poor"), with $T_1 + T_2 = T$. Let: $m_1 = \min(....,\mu_{t_1},....)$; $M_2 = \max(....,\mu_{t_2},....)$. The required j.m.f. for the particular pattern of interest is given by the following: [6]

$$\text{JMF} = \max(0, m_1 - M_2). \tag{6.9}$$

Different types of longitudinal measures correspond to, or can be simply derived from, different patterns S. A number of applications are described

[6] Note that this is the intersection of m.f.'s of opposite types, m_1 and $\overline{M}_2 = 1 - M_2$, using the Bounded operator $\text{JMF} = \max(0, m_1 + \overline{M}_2 - 1)$. Note also that throughout we use symbols μ to represent propensities of the same type (e.g. propensity to "poverty"); it is the type of cross-sectional sets of interest which are different in the two groups (e.g. S_1 "poor", S_2 "non-poor").

later. As an example, for the propensity to be poor at time 1, non-poor at time 2, and then re-entering poverty at time 3, we have:

$$S_1 = (1,3),\ S_2 = (2),\ JMF = \max\left(0, \min\left(\mu_1, \mu_3\right) - \mu_2\right). \qquad (6.10)$$

On the basis of the above, we formulate a general procedure in the following terms. Consider any sequence of cross-sectional propensities to poverty or deprivation. It can always be expressed in the form: $(\ldots, \mu_{t_1}, \ldots), (\ldots, \mu_{t_2}, \ldots)$, where t_1 indicates T_1 elements of the same type in one group, and t_2 indicates T_2 elements of the opposite type in the other group.

(i) Sort the elements into two groups by type, for instance all T_1 elements of one type followed by all T_2 elements of the other type.

(ii) Construct the intersection for each group involving elements of the same type using the Standard operator.

(iii) Finally, construct the intersection of the two results of the above operation using the Bounded operator (equation 6.9).

Since the temporal order of cross-sectional propensities is immaterial in the construction of their intersection using this rule, we may view the application of this rule as being without memory. More precisely perhaps, we may designate it as a procedure "without chronology": the outcome depends on the whole "history" (i.e., the specified type of cross-sectional sets in the time sequence t=1 to T, and the associated membership functions); but it does not depend on the actual chronology, the temporal sequence, of those cross-sections[7].

Marginal constraints

As noted, a set of T cross-sections yields 2^T longitudinal sequences. In the conventional analysis, these represent 2^T exhaustive and non-overlapping classes, with each individual unit belonging to one and only one of these, i.e. having some particular pattern (k) of poverty and non-poverty over the T years. Population totals or proportions over any grouping of these patterns are clearly additive. The same consistency must also hold under fuzzy conceptualization.

This condition is ensured by marginal constraints. The above procedure satisfies all the required marginal constraints (Betti et al. 2005b), as can be noted from the following. By definition, all the marginal constraints involved are expressed by successive applications of the following relationship:

[7] This procedure has certain similarities with that proposed by Betti, Cheli and Cambini (2004). However, the present procedure is more general and more consistent.

$$\mathrm{I}_{t-1} = \mathrm{I}_t + \bar{\mathrm{I}}_t \text{, t = T to 1, (with } \mathrm{I}_0 = 1) \tag{6.11}$$

Here I_t is the joint membership of an individual in a particular longitudinal sequence of length t. As before, let (S_1, S_2) represent the two groups of terms of opposing types in the sequence. I_t and $\bar{\mathrm{I}}_t$ differ only in that for one of the time periods in the sequence, the cross sectional sets considered are of opposite types. In other words, $\bar{\mathrm{I}}_t$ is the degree of joint membership for the sequence obtained from I_t by replacing any particular term in one of the groups (say S_1) with complement of that term in the other group (S_2). I_{t-1} is the degree of joint membership obtained by removing that term altogether, giving a sequence of only (t-1) terms. Equation (6.11) states that $(\mathrm{I}_t, \bar{\mathrm{I}}_t)$ are fuzzy partitions of I_{t-1}; that of course is exactly what is meant by "marginal constraints". It can be seen that $\bar{\mathrm{I}}_t$ is non-zero only if the term moved from S_1 is the smallest term in that set (otherwise, $\mathrm{I}_{t-1} = \mathrm{I}_t$). In this case, let $m_1^{(2)}$ be the second largest value in S_1 (i.e. the smallest value left after the move). It can be seen that:

$$I_t = \max(0, m_1 - M_2),$$

$$\bar{I}_t = \max\left(0, m_1^{(2)} - \max(m_1, M_2)\right),$$

$$I_{t-1} = \max\left(0, m_1^{(2)} - M_2\right),$$

which satisfies the required marginal constraint (6.11).

6.6 Application to specific situations

In this section we describe some important applications of the above rule for the construction of fuzzy intersections defining longitudinal measures.

6.6.1 Persistence of poverty

Analysis of the persistence of poverty over time requires the specification of j.m.f.'s of the type:

$$\mathrm{I}_\mathrm{T} = \mu_1 \cap \mu_2 \ldots\ldots\ldots \cap \mu_\mathrm{T} \text{ ,}$$

$$\mathrm{U}_\mathrm{T} = \mu_1 \cup \mu_2 \ldots\ldots\ldots \cup \mu_\mathrm{T} \text{ ,}$$

where the first expression is the intersection of a series of T cross-sectional m.f.'s for any individual unit, and the second expression is their union. I_T represents the individual's propensity to be poor at all T periods. U_T is the propensity to be poor at *at least one* of the T periods. Since all sets $\mu_1,.....,\mu_T$ are of the same type (all being propensities to "poverty" rather than to "non-poverty"), the Standard operations apply:

$$I_T = \min(\mu_1, \mu_2, ..., \mu_t, ...\mu_T)$$

$$U_T = \max(\mu_1, \mu_2, ..., \mu_t, ...\mu_T).$$

The complement of U_T, $\overline{U_T} = (1 - U_T)$, is the propensity to be *never poor*[8].

The propensity to be poor in exactly t out of T years is the sum of j.m.f.'s over all sequences with t cross-sectional sets of the type "poor" and the remaining (T–t) of the type "non-poor". For any particular sequence of this type, rearrange the sets such that the *first* t terms are of the type "poor". Hence the j.m.f. for the particular sequence is:

$$JMF = \max[0, \min(\mu_1, \mu_2, ..., \mu_t) - \max(\mu_{t+1}, \mu_{t+2}, ..., \mu_T)],$$

which is non-zero only for one sequence in which the first group contains the t largest memberships. With [t] denoting the ordered sequence of decreasing μ values, the required j.m.f. becomes:

Poor (exactly t out of T years): $\mu_{[t]} - \mu_{[t+1]}$,

and by simple addition:

Poor (at least t out of T years): $\mu_{[t]}$, the tth largest value.

If we define "persistent" poverty, for instance, as the propensity to be poor over at least a majority of the T years, i.e. over at least t years, with t=int(T/2)+1, the smallest integer being strictly larger than (T/2), the required propensity to persistent poverty is the $[\text{int}(T/2)+1]^{th}$ largest value in the sequence $(\mu_1,......\mu_T)$.

With the conventional poor/non-poor dichotomy, any individual spends some specified number of years between 0 and T in the state of poverty during the interval T. With poverty treated as a matter of degree, any particular individual is seen as contributing to the *whole distribution*, from 0 to T, of the number of years spent in poverty. Over an interval of T years

---

[8] The same result is obtained by considering intersection of non-poor sets:

$$\overline{U_T} = \min(\overline{\mu}_1, \overline{\mu}_2, ..., \overline{\mu}_t, ...\overline{\mu}_T) = 1 - \max(\mu_1, \mu_2, ..., \mu_T) = 1 - U_T.$$

the proportion of the time spent in poverty by the ith individual is (with $\mu_{[T+1]} = 0$):

$$t_i = \sum_{t=1}^{T} t \cdot \left(\mu_{[t]} - \mu_{[t+1]}\right) / T = \sum_{t=1}^{T} \mu_t \Big/ T,$$

i.e. simply the mean over the T periods of an individual's cross-sectional propensities to poverty.

6.6.2 Rates of exit and re-entry

Consider for instance the following. Given the state of poverty at time 1, and also at a later time (t–1), what is the proportion exiting from poverty at time t=2, 3, …? Given the state of poverty at time 1, but of non-poverty at a later time (t–1), what is the proportion which has re-entered poverty at time t=3, 4, ….?

In conventional analysis, the above rates are computed simply from the count of persons in various states. Consider for instance individuals poor at times t and (t–1). For exit rate at time t, the numerator is the count of persons poor at times 1 and (t–1), but non-poor at time t; the denominator is the count of all persons who are poor at times 1 and (t–1). Similarly for persons poor at time 1, non poor at (t–1) but poor again at t, the re-entry rate numerator is the count of persons poor at time 1, non-poor at time (t–1), but poor again at time t; the denominator is the count of persons who are poor at time 1 and non-poor at time (t–1). The construction of these measures using fuzzy m.f.'s is also straightforward. With μ_t as a person's propensity to poverty at time t, the person's contribution of these rates is as follows.

Exit rate:

Numerator $\left(\mu_1 \cap \mu_{t-1}\right) \cap \bar{\mu}_t = \max\left[0, \min\left(\mu_1, \mu_{t-1}\right) - \mu_t\right]$

Denominator $\left(\mu_1 \cap \mu_{t-1}\right) = \min\left(\mu_1, \mu_{t-1}\right)$.

Re-entry rate:

Numerator

$$\mu_1 \cap \bar{\mu}_{t-1} \cap \mu_t = \left(\mu_1 \cap \mu_t\right) \cap \bar{\mu}_{t-1} = \max\left[0, \min\left(\mu_1, \mu_t\right) - \mu_{t-1}\right]$$

Denominator $\mu_1 \cap \bar{\mu}_{t-1} = \max\left[0, \mu_1 - \mu_{t-1}\right]$.

The corresponding rates for the population are computed by simply averaging the above individual contributions.

6.7 Concluding remarks

When poverty is viewed as a matter of degree in contrast to the conventional poor/non-poor dichotomy, that is, as a fuzzy state, two additional aspects are introduced into the analysis.

(i) The choice of membership functions i.e. quantitative specification of individuals' or households' degrees of poverty and deprivation.

(ii) And the choice of rules for the manipulation of the resulting fuzzy sets, rules defining their complements, intersections, union and aggregation. Specifically, for longitudinal analysis of poverty using the fuzzy set approach, we need *joint membership functions* covering more than one time period, which have to be constructed on the basis of the series of cross-sectional *membership functions* over those time periods.

This Chapter has discussed approaches and procedures for constructing fuzzy measures of income poverty and of combining them with similarly constructed measures of non-monetary deprivation using the fuzzy set approach.

In fact, the procedures for combining fuzzy measures in multiple dimensions at a given time are identical, in formal terms, to the procedures for combining fuzzy cross-sectional measures over multiple time periods. We have proposed a general rule for the construction of fuzzy set intersections, that is, for the construction of a longitudinal poverty measure from a sequence of cross-sectional measures under fuzzy conceptualization. This general rule is meant to be applicable to any sequence of "poor" and "non-poor" sets, and it satisfies all the marginal constraints. On the basis of the results obtained, various fuzzy poverty measures over time can be constructed as consistent generalizations of the corresponding conventional (dichotomous) measures.

Numerical results of these procedures applied to measures of multidimensional poverty and deprivation, and to combinations of such measures have been presented elsewhere.

References

Betti G, Cheli B, Cambini R (2004) A statistical model for the dynamics between two fuzzy states: theory and an application to poverty analysis. Metron 62:391-411

Betti G, Cheli B, Lemmi A, Verma V (2005a) On the construction of fuzzy measures for the analysis of poverty and social exclusion. Presented at the International Conference in Memory of two Eminent Social Scientists: C. Gini and M.O. Lorenz, Siena, Italy, May 2005

Betti G, Cheli B, Lemmi A, Verma V (2005b) On longitudinal analysis of poverty conceptualised as a fuzzy state. Presented at the First Meeting of the Society for the Study of Economic Inequality (ECINEQ), Palma de Mallorca, Spain, July 2005

Betti G, Verma V (1999) Measuring the degree of poverty in a dynamic and comparative context: a multi-dimensional approach using fuzzy set theory. Proceedings ICCS-VI, Vol. 11, pp. 289-301, Lahore, Pakistan

Betti G, Verma V (2002) Non-monetary or Lifestyle Deprivation. In: Eurostat, European Social Statistics: Income, Poverty and Social Exclusion: 2nd Report, Luxembourg: Office for Official Publications of the European Communities, pp 76-92

Betti G, Verma V (2004) A methodology for the study of multi-dimensional and longitudinal aspects of poverty and deprivation. Università di Siena, Dipartimento di Metodi Quantitativi, Working Paper 49

Cerioli A, Zani S (1990) A Fuzzy Approach to the Measurement of Poverty. In: Dagum C, Zenga M (eds) Income and Wealth Distribution, Inequality and Poverty, Studies in Contemporary Economics, Springer Verlag, Berlin, pp 272-284

Cheli B (1995) Totally Fuzzy and Relative Measures in Dynamics Context. Metron 53:183-205

Cheli B, Betti G (1999) Fuzzy Analysis of Poverty Dynamics on an Italian Pseudo Panel, 1985-1994. Metron 57:83-103

Cheli B, Lemmi A (1995) A "Totally" Fuzzy and Relative Approach to the Multidimensional Analysis of Poverty. Economic Notes 24:115-134

Chiappero-Martinetti E (2000) A multidimensional assessment of well-being based on Sen's functioning approach. Rivista Internazionale di Scienze Sociali 108:207-239

Clark D, Qizilbash M (2002) Core poverty and extreme vulnerability in South Africa. The Economics Research Centre, School of Economic and Social Studies, University of East Anglia, Discussion Paper 2002/3

Eurostat (2002) European Social Statistics: Income, Poverty and Social Exclusion: 2nd Report. Luxembourg: Office for Official Publications of the European Communities

Hagenaars AJM (1986) The Perception of Poverty. North Holland, Amsterdam

Klir GJ, Yuan B (1995) Fuzzy Sets and Fuzzy Logic. Prentice Hall, New Jersey

Lelli S (2001) Factor Analysis vs. Fuzzy Sets Theory: Assessing the Influence of Different Techniques on Sen's Functioning Approach. Discussion Paper Series DPS 01.21, November 2001, Center for Economic Studies, Catholic University of Louvain

Maasoumi E, Nickelsburg G (1988) Multivariate measures of well-being and an analysis of inequality in the Michigan data. Journal of Business & Economic Statistics 6:326-334

Nolan B, Whelan CT (1996) Resources, deprivation and poverty. Clarendon Press, Oxford

Ram R (1982) Composite indices of physical quality of life, basic needs fulfillment and income. A principal component representation. Journal of Development Economics 11:227-248

Verma V, Betti G (2002) Longitudinal measures of income poverty and life-style deprivation. Università degli Studi di Padova, Dipartimento di Scienze di Statistiche, Working Paper 50

Whelan CT, Layte R, Maitre B, Nolan B (2001) Income, deprivation and economic strain: an analysis of the European Community Household Panel. European Sociological Review 17:357-372

Zadeh LA (1965) Fuzzy sets. Information and Control 8:338-353

7 French Poverty Measures using Fuzzy Set Approaches

Valérie Berenger, Franck Celestini

University of Nice-Sophia Antipolis

7.1 Introduction

Traditionally, poverty has been defined as a lack of income and has been associated with the study of personal income. Pareto was the first to analyse and estimate a model of income distribution. Although this finding has been largely disputed, it has opened a fruitful area of research consisting in analysing and identifying theoretical distribution functions associated with income.

Beside these studies, the poverty concept has considerably evolved during the last three decades. New definitions have emerged since the seminal works of Townsend (1979) and Sen (1985). They point out the limitations of income as a proxy of well-being and its dual poverty and the arbitrariness inherent in the identification of the poor according to a poverty line defined in reference to the mean or to the median income of the society. These new approaches underline the multidimensional and the vague aspects imbedded in the poverty concept and have led to the search for new methodological tools in order to deal with these two aspects. The fuzzy set theory offers a suitable mathematical tool for this purpose.

First applied by Cerioli and Zani (1990) in the context of multidimensional poverty analysis, it has given rise to theoretical refinements by Cheli and Lemmi (1995) who proposed a "Totally Fuzzy and Relative" procedure to the measurement of poverty called the TFR approach. Fuzzy set methodology has been recently used in several contributions which address the operationalization of Sen's Capability Approach (Chiappero-Martinetti 1996, 2000; Lelli 2001; Qizilbash 2002).

This methodology consists in extending the traditional definition of the membership to a given set. Instead of dividing the population between poor and non poor as is the case in the traditional money metric measurement of poverty, the fuzzy set theory presents the advantage of taking into account a continuum of situations between these two extremes. In contrast to the income based measure, the method allows for the use of data from census or surveys which provide useful information about various aspects of living conditions of households or individuals in order to go beyond the traditional view which determines poverty as a lack of command over resources and to overcome problems relative to the definition of a poverty line. This methodology is assumed to give a different picture of poverty. However, as in any multidimensional method of measurement, the TFR approach raises the question of the number of indicators to include for a robust poverty measure. As such a multidimensional procedure defines a poverty score or a degree of deprivation for each household, it would be relevant to perform an analysis of its distribution along the same lines as was the case for income in order to characterize the organization of poverty in a given society.

In Sect. 7.2, the TFR approach is applied to a vector of attributes that relate to the main basic relevant aspects of living conditions of households using data from the *French Surveys on Living Conditions* for the two different years 1986 and 1993. The results obtained are then compared to the ones produced from the income based poverty measurement. Sect. 7.3 considers the robustness of the fuzzy poverty index by carrying out a sensitivity analysis according to the number of considered variables. In Sect. 7.4, a new method based on the TFR approach is proposed. It is applied to the case of France in order to deal with the possibility of extracting a law from multidimensional poverty scores analogous to the power law identified by Pareto from income data. Finally, concluding comments are given in Sect. 7.5.

7.2 Application of the TFR approach using data from the *French Surveys on Living Conditions* for the years 1986 and 1993

We apply the methodology proposed by Cheli and Lemmi (1995) using two databases both coming from an *INSEE Survey of Living Conditions* and distributed by the *LASMAS-IDL CNRS*. The size of the household sample is 13154 for the year 1986 and 13280 for the year 1993. This methodology consists of three steps.

The first one relates to the definition of poverty. In order to assess deprivation relative to the living conditions of the sampling population, we need to select a vector of v attributes from the survey. The selection can be constrained by the availability of the data or be dependent on the purpose of the analysis. It can also be made using subjective perceptions of individuals (Mack and Lansley 1985). Whatever the selection, it is subject to arbitrariness and leads to the question of the nature and the number of dimensions to retain. Here, we select k = 17 variables relative to the ownership of durable goods, the housing characteristics, the quality of the housing and the ownership of assets. These indicators do not cover the full range of variables that are relevant or available. The choice is due to their relative easiness of use in conjunction with fuzzy measures. Following Cheli and Lemmi, these attributes can be interpreted as indicators of effects that reflect symptoms of insufficient resources. In the framework of Sen's capability approach, they can be viewed as the opportunities provided by income in order to have access to good living conditions in terms of housing, durable goods and assets. As it can be observed, variables are of various types. Income is not considered in this selection as insufficient incomes can intervene as one of various explanatory variables in some dimensions like housing conditions, for example. For the two years considered, the same attributes or proxies have been retained. The full list is given in the appendix.

The second step of the methodology provides a measure of the degree of deprivation relative to a given attribute for each household or individual. A sampling of n households and a vector of k indicators j is considered. For each indicator, a function $x_j^{(l)}$ is defined with m = 1 … M, the possible values of modalities taken by j rearranged by increasing order, where higher values denote a higher risk of poverty. The value $x_j^{(1)} = 0$ corresponds to the lowest risk of poverty and $x_j^{(M)} = 1$ to the highest risk of poverty associated with the deprivation indicator j.

In contrast to the Cerioli and Zani approach, Cheli and Lemmi propose that the degree of poverty relative to indicator j should be directly proportional to the cumulative distribution function:

$$F\left(x_j\right) = \int_0^{x_j} f_j\left(x\right) dx \tag{7.1}$$

This assumption is based on the fact that the poverty feeling of a household is directly related to the number of households owning a good that it does not own itself. In other words, this approach stresses and takes into account the relative nature of the poverty feeling. Within the TFR approach, the degree of poverty or equivalently, the value of the membership function to the subset of poor, $\mu_j(i)$ of the i-*th* household with regard to indicator j satisfies the following specification:

$$\mu_j\left(x_{ij}\right) = \begin{cases} 1 & if \quad x_{ij} = x_j^{(s)} \\ \mu_j\left(x_j^{(l-1)}\right) + \dfrac{F(x_j^{(l)}) - F(x_j^{(l-1)})}{1 - F(x_j^{(l)})} & if \quad x_{ij} = x_j^{(l)} \\ 0 & if \quad x_{ij} = x_j^{(1)} \end{cases} \tag{7.2}$$

where x_{ij} is the value taken by indicator j for the i-*th* household and $\mu_j(x_j^{(l-1)})$ represents the degree of deprivation attached to the modality (l–1) for indicator j. With this formulation, the degree of deprivation $\mu_j(x_{ij})$ lies in the interval [0,1] and increases with the risk of poverty.

Finally, the degrees of poverty assessed according to each of the k deprivations need to be reduced to one dimension in order to obtain a multidimensional poverty index $\mu(i)$ for each household i. The poverty index $\mu(x_i)$ is then defined as the weighted average with respect to the k indicators:

$$\mu(x_i) = \frac{\sum_{j=1}^{k} \mu(x_{ij}) w_j}{\sum_{j=1}^{k} w_j} \tag{7.3}$$

where w_j is the weight associated to the j-*th* indicator. For the weighting system, Cheli and Lemmi (1995) propose the following expression:

$$w_j = \ln\left(\frac{1}{\overline{\mu}_j}\right) \Big/ \sum_{j=1}^{k} \ln\left(\frac{1}{\overline{\mu}_j}\right) \tag{7.4}$$

The weight w_j is chosen as an inverse function of the average degree of poverty: $\overline{\mu}(x_j) = \frac{1}{n}\sum_{i=1}^{n} \mu(x_{ij})$. This means that a considerable weight is given to a variable j associated to a very widespread good in the society. In other words, the higher the ownership of the good, the poorer the household that does not own this good. Finally, a global index of poverty for the overall population considered can be derived which is defined by the average of individual deprivation indices:

$$P = \frac{1}{n}\sum_{i=1}^{n} \mu(x_i) \tag{7.5}$$

The value of P represents the proportion of households belonging in a fuzzy sense to the poor subset.

Table 7.1 gives the results of the application of this methodology using the attributes selected from the *French Surveys on Living Conditions* for the years 1986 and 1993 attached respectively to n = 13154 and n = 13280 households.

According to the TFR approach, it appears that 14.15 % and 13.4 % of the households are poor respectively in 1986 and 1993 with a relatively low decrease between the two years. Although TFR does not possess a poverty line dividing the population into poor and non poor, it is possible to adopt a dichotomous approach in order to derive a critical value for the poverty score from the cumulative distribution of poverty indices and in particular from the global poverty index (P). This value indicates that any household whose poverty score is greater than this critical value is considered as poor.

Table 7.1. Summary statistics on multidimensional TFR index of poverty

	1986	1993
Value of P	0.1415	0.1348
Standard deviation	0.1288	0.1198
Critical value of poverty index	0.275	0.259

Since the TFR index of poverty is defined as a function of the main basic indicators of living conditions of households, it is intended to give a different picture of poverty than the one based on income as single indicator. Nevertheless, as the selected indicators may be considered as representative of insufficient economic resources, a similarity between the two approaches is possible. In order to see if the two different approaches indicate the same subsets of poor households, the equivalent income I(i) of the i-*th* household is extracted from the same sample survey used to establish poverty scores. It is the total income obtained from the sum of all the individuals of the household adjusted with the OECD equivalence scale.

Using the value of the fuzzy global poverty index P, we can select a first subset of households being poor according to the TFR approach. Its size is of $n*P = 1860$ for the year 1986. We then consider a second subset of households of the same size reporting the smallest incomes. From this selection, it appears that only 542 households belong to the two subsets. We can easily infer[1] that if the two subsets were completely uncorrelated then the number of common households that one could find would be equal to $n*P^2 = 263$.

As it appears, the multidimensional measure provides a different picture of poverty. The overlap between the two measures is relatively small.

[1] For more details, see Berenger and Celestini (2004). In this contribution, carrying out a statistical comparison between multidimensional and unidimensional measures of poverty, an agreement coefficient has been defined using a free parameter identified to the poverty line in unidimensional approaches. The agreement coefficient gives a measure of the correspondence between the subsets of poor according to the income based poverty measure and the multidimensional one.

7.3 Statistical sensitivity analysis of the TFR poverty index on the number of attributes

According to the TFR approach, the index P belongs to the class of additively decomposable indices. It can be broken down into subgroups of households and subgroups of attributes in order to exploit information from the multidimensional approach to poverty and to help to identify the main causes of poverty. As pointed out by Miceli (1998), at this stage of the analysis, the interpretation must be made with caution given the different types of variables defining dimensions of poverty. Regardless of this aspect, another problem that arises in any multidimensional approach concerns the number of dimensions or attributes to retain. In the context of the TFR approach, we may wonder if the relevance of the results obtained for the value of P is sensitive to the number of attributes included in its evaluation. In answer to this, performing a sensitivity analysis of this dependence is proposed. We use the empirical study above based on the selection of k = 17 attributes and we consider the values of P obtained successively for a selection of v = 2, 3,...17 variables. In order to avoid the dependence between the computed values of P and the rank of variables selected in the list, all possible subsets of v = 2,3...17 variables that can be extracted from the k = 17 variables are taken into account. For a k-subset of attributes, the number of possible combinations is equal to:

$$C_v^k = \frac{k!}{v!(k-v)!} \tag{7.6}$$

For each v-subset of attributes, the application of the TFR method provides distributions of various values for the TFR poverty index P and their derived standard deviations. These results are summarized by averaging them over all possible subsets of v variables. Figure 7.1 represents mean values of the TFR poverty index as a function of the number of variables v for the two years considered.

As we can observe, the mean value of P is a decreasing function of the v-subset. It can be tentatively explained by the fact that when considering an increasing number of indicators of living conditions, the probability of being identified as poor tends to be much less than when a low number of aspects of living conditions are retained. It also appears that the value of P tends to reach a stable value from v-subsets including roughly more than 10 variables. According to the sensitivity. analysis, the robustness of the poverty measure based on the TFR approach appears to be dependent on the number of indicators included in the definition of multidimensional poverty. Relevant empirical measures can only be obtained when a sufficient number of indicators are considered. According to this result, the relevance of a decomposition of the TFR poverty index according to each dimension or subgroups of attributes would deserve to be submitted to such a sensitivity analysis.

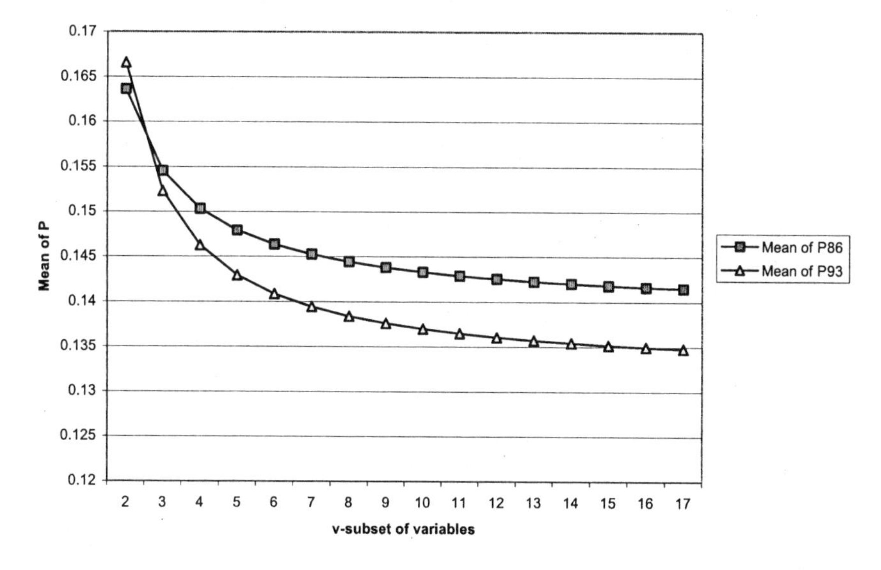

Fig. 7.1. Mean TFR poverty index as a function of v-subset of variables

7.4 Extracting a law from multidimensional poverty scores analogous to the Pareto Law for income distribution: a method based on the TFR approach

Although the TFR approach provides an alternative to the traditional analysis of poverty, it is worth remembering that traditionally, poverty has been associated with the analysis of income distribution. The starting point of these studies originates with Pareto who was the first to demonstrate that the distribution of personal income was best fitted by a power law, the parameter of which could be interpreted as an index of inequality. Although this finding has been largely disputed, it has opened a fruitful area of research among statisticians, mathematicians and probability theorists. The models frequently used are Pareto, log-normal and gamma characterized by two parameters[2]. Recently, this topic has attracted the interest of econophysicians. Among the latter, Dragulescu and Yakovenko (2000) es-

[2] For more details, see Dagum (1990, 1999) and Kakwani (1980b).

tablished that income would follow an exponential law for the great majority of the population in the UK and in individual states of the USA.

Surprisingly, as for income, no attempt has been made in the context of multivariate analysis of poverty to characterize the distribution of poverty scores by an appropriate theoretical density function. Such an attempt would permit the use of graphic devices in order to provide a more revealing picture of the degree of poverty relative to each household and its distribution. Such a device would allow for the detection of possible laws from data as well as obtaining some information on the organization of poverty in a given society. The study of functional distribution of multidimensional poverty scores could have useful applications in poverty comparisons across times, regions and countries.

In order to study the distribution of multidimensional poverty scores, it is necessary to define a method that provides poverty scores lying, as for income, between 0 and infinity. The proposed method is derived from the TFR approach.

We consider $i \in [1, n]$ households and $j \in [1, k]$ indicators selected above. According to the formulation of the membership function given by (7.2), the degree of deprivation $\mu(x_{ij})$ lies in the interval [0,1] and increases with risk of poverty. However, as is the case for income or wealth, we need to introduce a poverty score that is not limited but naturally lies between 0 and infinity. For this purpose, the following new definition of $\mu(x_{ij})$ is proposed:

$$s(x_{ij}) = \mu(x_{ij}) = \ln\left(\frac{1}{1-F(x_{ij})}\right) \tag{7.7}$$

where $s(x_{ij})$ is still an increasing function of $F(x_{ij})$ but is no longer restricted to the interval [0,1]. Even if numerous alternative expressions can be proposed, this one seems to be the simplest and introduces a logarithmic function often present in the measure of human sensitivities. The degree of poverty of the i-*th* household s(i) is then defined as the weighted average with respect to the k indicators:

$$s(i) = \sum_{j=1}^{k} w_j s(x_{ij}) \tag{7.8}$$

with w_j the weight of the j-*th* indicator.

As in the TFR approach, the weight w_j is chosen to be an inverse function of the average of degree of poverty. The idea is of the same type as the one used above for justifying the proportionality between $s(x_{ij})$ and the cumulative distribution $F(x_{ij})$. The weights are henceforward represented with the following expression:

$$w_j = \ln\left(1+\frac{1}{\bar{s}_j}\right) / \sum_j \ln\left(1+\frac{1}{\bar{s}_j}\right) \tag{7.9}$$

This expression satisfies the inverse relation between the weight and the mean score and uses another logarithmic function. The form $\ln\left(1+\frac{1}{\bar{s}_j}\right)$ is chosen to prevent the occurrence of negative weights. Indeed, unlike the classical TFR approach, scores can be greater than 1. The denominator ensures the normalisation of the weights, avoiding a trivial dependence of s(i) with k. Starting from the x_{ij} functions, a multidimensional poverty score s(i) is evaluated for each of the n households.

Applying this method to the case of France for the years 1986 and 1993, the poverty scores and the associated probability density function ρ(s) are computed. In Figure 7.2, the direct measurement of the probability function of poverty score for the year 1986 is given. Circles correspond to empirical data and the full line is an exponential law fit to the data. We can see that the empirical distribution is rather well fitted by an exponential distribution. However, for practical reasons, the empirical probability distribution function does not give an accurate description at low poverty score levels.

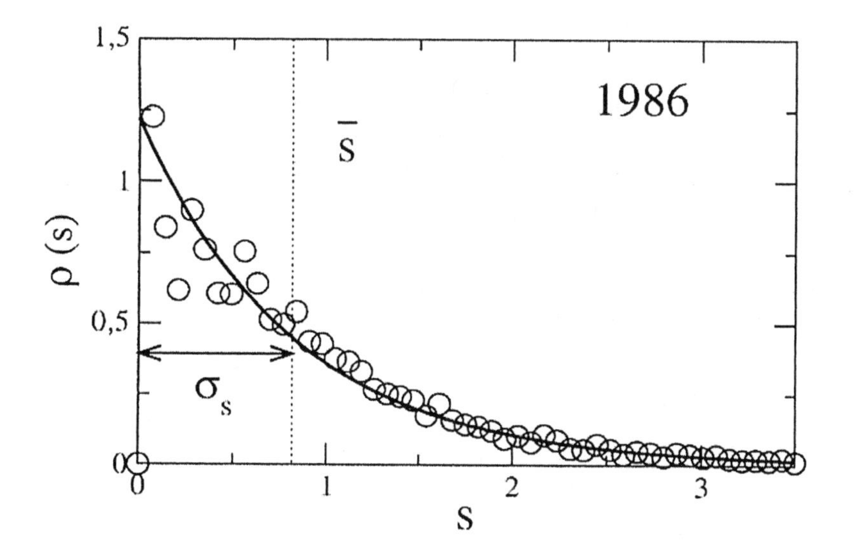

Fig. 7.2. Probability density function of poverty scores for the year 1986

In order to improve the identification of the nature of ρ(s), we apply the "rank ordering method" frequently used in the analysis of income and wealth distributions. This technique is very close to the construction of a cumulative distribution. It allows the adjustment of the statistical fluctuations that affect the cumulative distribution at fixed score by reporting them on the variable itself at fixed rank i. The n poverty scores are reordered by decreasing values: $s_{[1]}$ and $s_{[n]}$ are respectively the largest and the smallest values. The rank ordering method consists in identifying the relation between the i-*th* larger value $s_{[i]}$ and its rank i.

Figure 7.3 represents the $s_{[i]}$ values as a function of their rank i for the years 1986 and 1993. As is well known, for an exponential probability function of the form:

$$\rho(s) = \frac{1}{\sigma_s} \exp\left(-s/\sigma_s\right) \tag{7.10}$$

the i-*th* larger value of s satisfies:

$$s_{[i]} = -\sigma_s \ln\left(\frac{i}{n}\right) \tag{7.11}$$

Using a semi-logarithmic representation, the data points are well fitted by two different straight lines for the two different years.

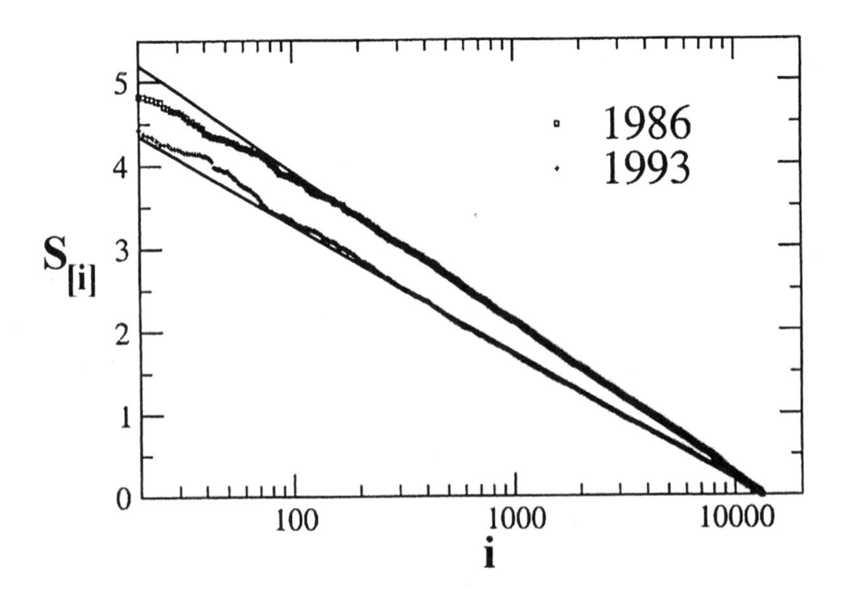

Fig. 7.3. Rank ordering of multidimensional poverty scores with logarithmic horizontal scale for the years 1986 and 1993

As we can observe, except for roughly the 100-th largest scores, the agreement between computed scores and the fitting functions is good. More than 99 % of the household scores follow the exponential distribution. This finding demonstrates the relevance of our computed scores because they are distributed according to a well-defined distribution law which outlines the way the society organizes itself in relation to this poverty score. The exponential distribution is a function characterized by a single parameter σ_s. Its mean is equal to its standard deviation and is also proportional to its median s_m:

$$\bar{s} = \sigma_s = s_m \ln(2)^{-1} \tag{7.12}$$

Taking into account the dependence between the multivariate measure of poverty and the number of attributes evidenced above, these three parameters are plotted as a function of the number of indicators (k) accounted for in the measurement of the multidimensional poverty score attached to each household. The distribution is computed using the method described, except that the analysis is restricted to the v first indicators without taking into account all the possible v-subsets of variables that can be extracted from the k = 17 variables.

As shown in Figure 7.4, the three parameters are rather different when few numbers of indicators are considered. As v increases, they assume closer and closer values. Finally for a number of variables roughly greater than 13, the three parameters reach a constant and almost equal value. The same comments are valid for the other year considered. This plot confirms the hypothesis of an exponential distribution of poverty and also demonstrates that a sufficient number of attributes needs to be taken into account in order to reach this distribution. For a low number of indicators, the statistic is too poor to clearly identify the exponential distribution. This result should appear in other multidimensional poverty measures.

The main property of this organization type is that we can fully characterize poverty of the sample society by a single parameter. Indeed, as stressed above, the exponential distribution is defined by the unique parameter σ_s as illustrated in Figure 7.2 for the year 1986.

All other well-known distributions have at least two parameters which characterize the different moments. At this stage, it is important to note how our multidimensional poverty scores lead to one of the simplest probability density functions. In this case, the knowledge of σ_s is sufficient for obtaining all other possible indicators from the distribution density function. As a consequence, the comparison of poverty scores across societies or across a chosen period is straightforward. One has just to compare the different values obtained for σ_s. This comparison study is usually not so easy because the underlying distribution density is not as simple as the exponential. In these cases, methods like stochastic dominance are required to test if the same ranking of the distribution is obtained whatever the index of poverty chosen.

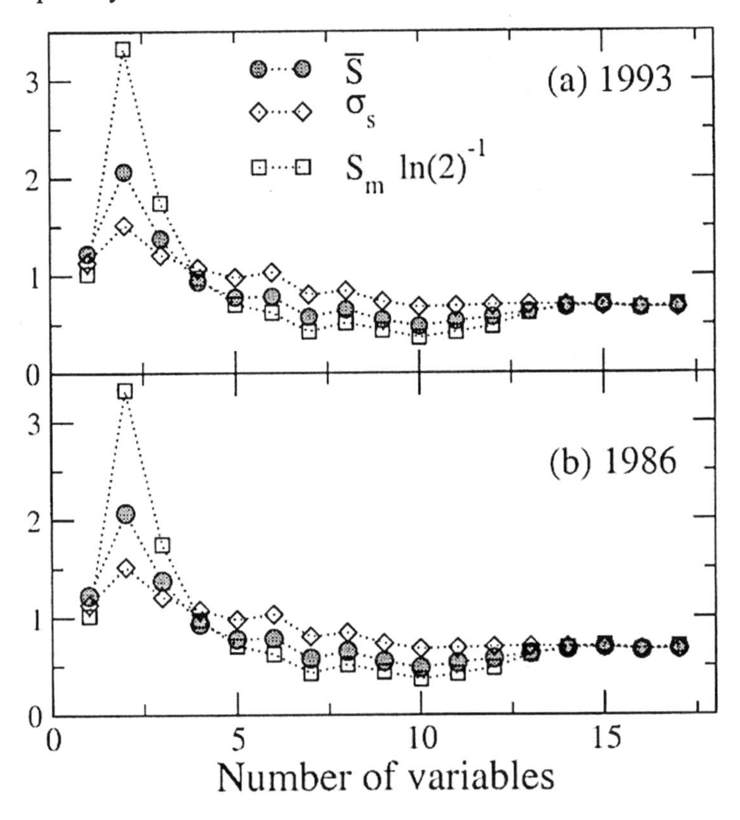

Fig. 7.4. Simple statistics of multidimensional poverty scores as a function of the number of attributes defining it.

For the case of France, the best fit to the household poverty scores for the two years gives the following values for σ_s: $\sigma_s = 0.819$ for 1986 and $\sigma_s = 0.672$ for 1993 which means a relative decrease of the mean poverty score of 17,9%. According to the properties of exponential distribution functions, the same relative variation stands for standard deviation and higher moments.

This finding can be used to predict how a global change in the poverty situation would be felt by different households of the society.

Let us assume that at time t the poverty distribution is characterised by an exponential distribution with a mean value of σ_{st} while at time t + dt the mean value decreases to $\sigma_{st} + d\sigma_{st}$. The relative poverty decrease is then defined as $r = d\sigma_{st} / \sigma_{st}$. It follows that when analysing the effect of a global policy without considering any individual increase or decrease in the poverty rank between t and t + dt, we can easily see that the relative individual poverty score variation is the same for each household and is equal to the mean relative variation (r). This result is directly connected to the exponential nature of the distribution. Conversely, considering a power law distribution, a mean variation would not be equally distributed among each individual or household. In that sense, the exponential distribution can be viewed as an egalitarian density distribution.

7.5 Concluding comments

In this Chapter, using data from the *French Surveys on Living Conditions* for the years 1986 and 1993, we have shown that the TFR approach offers a convenient tool for measuring poverty as a function of various variables that represent any aspect of living conditions. Although, the indicators selected in our application can be viewed as indicators of insufficient economic resources, the comparison analysis between households' fuzzy deprivation score and their equivalent income allows us to show that households defined as poor according to the TFR are not necessarily those reporting the smallest incomes. In other words, bad living conditions are not necessarily and uniquely associated to a lack of income. More important are the results obtained from the sensitivity analysis of fuzzy poverty index to the number of variables. A robust measure of the fuzzy poverty index can be obtained only when a sufficient number of indicators (roughly more than 10 in our case study) are accounted in its measurement. We have also considered how to extend the TFR approach in order to deal with the possibility of extracting a law from the distribution of multidimensional poverty scores analogous to the Pareto Law for income distribution. We proposed adjusting the TFR approach in order to define a poverty score lying between 0 and infinity. Applying the method to the case of France for the two years considered, we have found that the multidimensional poverty score is distributed according to an exponential law for almost all the population in consideration. The main property of this organization type is that we can fully characterize poverty of a sample society by a single parameter. As a consequence, the comparison of poverty in a chosen period is straightforward. One has just to compare the values of the parameter. This promising finding suggests examining the validity of the exponential law of the multidimensional poverty score for the whole population in all countries and at all times. The method could be applied using other vectors of indicators and data from different countries.

using other vectors of indicators and data from different countries. As is the case for income, other distributions of poverty scores like power law or exponential law with possible breakdown could be extracted from the data. Other possible distributions could evidence the validity of the exponential law to a certain range of poverty scores and the heterogeneity in the organization of poverty.

References

Berenger V, Celestini F (2004) Is There a Clearly Identifiable Distribution Function of Individual Poverty Scores? Presented at the 4th conference on the Capability Approach: Enhancing Human Security. University of Pavia, September 2004

Boettcher H (1994) The Use of Fuzzy Sets Techniques in the Context of Welfare Decisions. In: Eichhorn W (ed) Models and Measurement of Welfare Inequality. Springer, Heidelberg, pp 891-899

Cerioli A, Zani S (1990) A Fuzzy Approach to the Measurement of Poverty. In: Dagum C, Zenga M (eds) Income and Wealth Distribution, Inequality and Poverty. Springer, Heidelberg, pp 272-284

Cheli B, Ghellini G, Lemmi A, Pannuzi N (1994) Measuring Poverty in Transition via TFR Method: the Case of Poland in 1990-1991. Statistics in Transition 5:585-636

Cheli B, Lemmi A (1995) A "Totally" Fuzzy and Relative Approach to the Multidimensional Analysis of Poverty. Economic Notes 24:115–134

Chiappero Martinetti E (1996) Standard of Living Evaluation base on Sen's Approach: Some Methodological Suggestions. Notizie di Politeia 12:37-53

Chiappero Martinetti E (2000) A multidimensional Assessment of Well-Being Based on Sen's Functioning Approach. Rivista Internazionale di Scienze Sociali 108:207-239

Costa M (2003) A Comparison between Unidimensional and Multidimensional Approaches to the Measurement of Poverty. IRISS Working Papers Series 2003-02

Dagum C (1990) Generation and Properties of Income Distribution Functions. In: Dagum C, Zenga M (eds) Income and Wealth Distribution, Inequality and Poverty. Springer, Heidelberg, pp 1-17

Dagum C (1999) Linking the Functional and Personal Distributions of Income. In: Silber J (ed) Handbook on Income Inequality Measurement. Kluwer Academic Publishers, Boston, pp 101-132

Dagum C (2002) Analysis and Measurement of Poverty and Social Exclusion using Fuzzy Set Theory: Applications and Policy Implications. Working Paper, University of Bologna

Dragulescu A, Yakovenko VM (2000) Evidence for Exponential Distribution of Income in the USA. The European Physical Journal B 20:585-589

Dubois D, Prade H (1980) Fuzzy Sets and Systems: Theory and Applications. Academic Press, Boston
INSEE Enquête des Conditions de Vie des Ménages. Distributed by LASMAS-Idl C.N.R.S. (1986-1987), (1993-1994)
Kakwani N (1980a), On a Class of Poverty Measures. Econometrica 48:437-446
Kakwani N (1980b), Income Inequality and Poverty: Methods of Estimation and Applications. Oxford University Press, New York
Lelli S (2001) Factor Analysis vs. Fuzzy Sets theory: Assessing the influence of Different Techniques on Sen's Functioning Approach. Presented at the conference on "Justice and poverty: Examining Sen's Capability Approach". St. Edmund's College, Cambridge, June 2001
Mack J, Lansley S (1985) Poor Britain. Allen & Unwin, London
Miceli D (1998) Measuring Poverty using Fuzzy Sets. NATSEM, Discussion paper n°38
Pareto V. (1897) Cours d'Economie Politique. In: Bousquet GH, Busino G (eds), Geneva, Librairie Droz, 1965; vol.3
Qizilbash M (2002) A Note on the Measurement of Poverty and Vulnerability in the South African Context. Journal of International Development 14:757-772
Qizilbash M (2004) On the Arbitrariness and Robustness of Multi-Dimensional Poverty Rankings. WIDER Research Paper, 37
Sen A (1985) Commodities and Capabilities. Oxford University Press, Oxford India Paperbacks
Sen A (1992) Inequality Reexamined. Harvard University Press, New Delhi
Silber J (1999) Handbook on Income Inequality Measurement. Kluwer Academic publishers, Boston
Sornette D (2000) Critical Phenomena in Natural Sciences. Springer-Verlag, Berlin
Zadeh LA (1965) Fuzzy Sets. Information and Control 8:338-353

Appendix: List of deprivation indicators selected from the *INSEE-French Surveys of Living Conditions* 1986 and 1993[3].

- Television: $x_{1\ (m=3)}$
- Number of rooms in the accommodation: x_2 is a function of the ratio of the number of rooms per individual using household size
- Telephone in the accommodation: $x_{3\ (m=2)}$
- Water in the accommodation: $x_{4\ (m=3)}$
- Bathroom in the accommodation: $x_{5\ (m=5)}$
- Toilets in the accommodation: $x_{6\ (m=3)}$
- Kitchen in the accommodation: $x_{7\ (m=3)}$

[3] Number of modalities taken by each attribute appears in brackets.

- Owning a washing machine: $x_{8\ (m=3)}$
- Difficulty to heat the accommodation adequately (1986) /Bad insulation (1993): $x_{9\ (m=2)}$
- Humidity in the house: $x_{10\ (m=2)}$
- Environmental problems such as pollution and inadequate green space (1986)/Existence of bad smells (1993): $x_{11\ (m=2)}$
- Perception of the household situation according to the general aspect of the house: $x_{12\ (m=4 \text{ for } 1993 \text{ and } m=6 \text{ for } 1986)}$
- Noise around the house: $x_{13\ (m=2)}$
- Owning a car: $x_{14\ (m=2)}$
- House owned or rented: $x_{15\ (m=3)}$
- Banking account cheques: $x_{16\ (m=2)}$
- Cut of electricity, water and gas: $x_{17\ (m=3)}$
- Household total income and social transfers divided by the corresponding value of OECD equivalence scale (not included in our multidimensional measures).

8 The "Fuzzy Set" Approach to Multidimensional Poverty Analysis: Using the Shapley Decomposition to Analyze the Determinants of Poverty in Israel

Joseph Deutsch, Jacques Silber

Department of Economics, Bar-Ilan University

8.1 Introduction

The theory of fuzzy sets that originates in the work of Zadeh (1965) has been used in numerous fields during the past forty years and its applications cover fields as different as artificial intelligence, the stock market and poverty. The pathbreaking work of Cerioli and Zani (1990) has in fact launched a whole series of studies on the measurement of multidimensional poverty that are based on the idea of fuzzy sets. The specific contribution of the present Chapter is that it offers an illustration of the fuzzy set approach to poverty measurement based on Israeli Census data. It makes a systematic comparison of three fuzzy set approaches, the Totally Fuzzy and Absolute (TFA) approach of Cerioli and Zani (1990), the Totally Fuzzy and Relative approach (TFR) of Cheli and Lemmi (1995) and the Vero and Werquin (1997) approach (VW) and it finally shows how the so-called Shapley decomposition (see, Shorrocks 1999) may be used to find out which are the really important determinants of (multidimensional) poverty.

The chapter is organized as follows. The next section quickly summarizes the Vero and Werquin approach that is probably less known to the readers. Then various cross-tables are presented which show the impact of various factors on poverty. Such an analysis is then extended to a multivariate study of the determinants of multidimensional poverty based on logit analysis. The final section then uses the so-called Shapley procedure to find out which of these determinants are really important. Concluding comments are given at the end.

8.2 Theoretical Background

As mentioned previously three "Fuzzy Set" approaches to the measurement of multidimensional poverty will be examined in this chapter: the Totally Fuzzy and Absolute approach (TFA) originally suggested by Cerioli and Zani (1990), the Totally Fuzzy and Relative approach (TFR) suggested by Cheli and Lemmi (1995) and the approach (VW) suggested by Vero and Werquin (1997). Since the TFA and TFR approaches have been presented in previous chapters some information is only given on the VW approach.

One of the serious problems one faces when adopting a multidimensional approach for poverty measurement, such as the fuzzy approach, is that some of the indicators one uses may be highly correlated. To solve this problem, Vero and Werquin (1997) have proposed the following solution.

Let k be the number of indicators, n the number of individuals and f_i the proportion of individuals who are at least as poor as individual i when taking into account all the indicators. The following example illustrates the computation of f_i and is borrowed from Vero and Werquin (2002).

Let X_1 be equal to 1 if the household does not have a bathroom, X_2 be equal to 1 if the household does not have a car and X_3 be equal to 1 if the household does not have a phone. Let there be 6 individuals. Table 8.1 below gives the values assumed by these three indicators for each of them as well as the value of the indicator f_i.

Table 8.1. Computing the indicator f_i – a hypothetical illustration

Individual	X_1	X_2	X_3	Indicator f_i
1	0	1	1	4/6
2	1	1	1	1/6
3	0	1	1	4/6
4	0	0	0	6/6
5	0	1	1	4/6
6	1	0	0	2/6

Table 8.1 clearly shows that the (perfect) correlation between vectors X_2 and X_3 has been neutralized since the same values of f_i would have been obtained on the basis of vectors X_1 and X_2 or X_1 and X_3.

The deprivation indicator $m_P(i)$ for individual i will then be defined as:

$$m_P(i) = \frac{\ln\left(\frac{1}{f_i}\right)}{\sum_1^n \ln\left(\frac{1}{f_i}\right)} \quad if \ 0 < f_i \leq 1 \tag{8.1}$$

The membership function $\mu_P(i)$ for individual i is then expressed as:

$$\mu_P(i) = \frac{m_P(i) - Min[m_P(i)]}{Max[m_P(i)] - Min[m_P(i)]} \tag{8.2}$$

Finally the average value of the membership function P, over all individuals, is, as for the TFA and TFR approaches, defined as:

$$P = \frac{1}{n}\sum_{i=1}^{n} \mu_P(i) \tag{8.3}$$

8.3 The Case of Israel in 1995

The empirical illustration is based on data from the last Census that was carried out in Israel in 1995. Let us first describe the variables that were selected to derive estimates of the degree of multidimensional poverty in Israel during that year.

8.3.1 Selecting the Indicators

When preparing such a list of indicators it is impossible to ignore the type of society that is analyzed. Central heating for example may be relevant in most parts of Western Europe but is of much less interest in most parts of countries like Israel that are characterized by a kind of a subtropical climate.

Another important issue concerns the selection of the total number of indicators one wishes to take into account. Some information may quickly become redundant here. As was stressed in the illustration of Table 1 the ownership of one durable good may be very highly correlated with that of another durable good. The estimation technique proposed by Vero and Werquin (1997) and mentioned earlier tries in a way to solve this problem.

Another conceptual issue concerns the fact that some households may freely decide not to want to own certain types of durables. Thus, in Israel, some extremely religious households never buy a television set because it would distract their attention from more spiritual activities and might even have "perverse" effects on the education of the children.

Another question concerns the evaluation of the quality of a durable good. "Owning a car" may refer to very different situations, depending on whether it is one that has been bought as third or fourth hand and is small or a new, powerful and big car. It might therefore be useful, when estimating the degree of "fuzzy poverty", to include, whenever it is possible, whatever information is available on the quality of the item.

8.3.2 The Data Sources

Our source of information was the 1995 Israeli Census that includes only information available on the ownership of durable goods. Although in many cases the available information was binary, some variables were polytomous. This was the case of the variables indicating the period in which the apartment or house was built, the number of rooms in the dwelling and even whether there was a bath or a shower in the dwelling (see, the Appendix for the exact listing of the categories distinguished). There was even a purely quantitative variable, the one giving the number of cars available for household use. Moreover we have actually used as indicator not the number of rooms or cars in the household but the number of rooms or cars per individual. Note that the ownership of the dwelling is defined as a dichotomous variable taking the value one when the dwelling is owned by the household and the value zero otherwise.

The list of variables and the values they can assume are given in the Appendix.

8.3.3 Computing the percentage of poor according to the various approaches

For the TFA approach in the case of polytomous variables we gave the values of 0 and 1 to the extreme cases while the values taken in the other cases were derived by linear interpolation, as is usually done in this approach. Let us now present the results that are summarized in Table 8.2, for each of the three "fuzzy approaches".

It appears that according to the Totally Fuzzy Absolute approach (TFA) originally proposed by Cerioli and Zani (1990) 23.9% of the households are poor. When the Totally Fuzzy and Relative approach (TFR) is adopted (see, Cheli and Lemmi, 1995) 26.4% of the households are poor while with the Vero-Werquin Approach (VW) (1997) 27.5% of the households are poor. Note also that in all three cases the median is smaller than the arithmetic mean, implying that the three distributions of the membership function are asymmetric.

Table 8.2. Summary indices for the membership function and percentage of poor households (arithmetic mean of the membership function)

Summary Indices for the membership function	FTA Approach	TFR Approach	VW Approach
Arithmetic mean (percentage of poor)	23.9	26.4	27.5
Standard deviation	16.8	17.3	12.8
Median	19.7	22.5	26.1

In Table 8.3 some information is given on the degree of overlapping between the three distributions of the membership function that have been distinguished (according to the TFA, TFR and VW Approach). It appears that 18.5% of all the households are poor according to the three approaches. 25.6% according to at least two approaches and 33.7% according to at least one approach. Even though there is a good amount of overlapping between the three distributions, they are far from identifying the same households as poor.

Table 8.3. Percentage of Poor Households. Degree of Overlapping between the Various Approaches

Households identified as poor by	Percentage of Households that are poor
all three indices	18.5
at least two indices	25.6
at least one index	33.7

8.3.4 The Determinants of multi-dimensional poverty

In this section we investigate the impact of various variables such as the gender or the age of the head of the household on the probability of a household being considered as poor. For each variable considered we examine the potential existence of such a link but obviously only simple correlations are looked at. A more sophisticated analysis looking at the marginal impact of a given variable, ceteris paribus, will be carried out later on when results of logit type regressions are presented.

Simple correlations

The role of the gender

Table 8.4 examines the effect of the gender of the head of the household. It appears that all three approaches indicate that female-headed households are more likely to be considered as poor than male-headed households.

The Role of the Size of the Household

In Table 8.5 the impact of the size of the household is examined. It appears that whatever the approach adopted there is a U-shape relationship between the size of the household and the probability that it is poor. According to both the Totally Fuzzy Absolute and the Totally Fuzzy and Relative approaches, the probability of being considered as a poor household is smallest for households including five individuals. According to the Vero-Werquin Approach this probability is smallest for households with four members. As expected, in all three cases, the probability for a household to be poor is highest when it includes at least 10 members. With the Vero-Werquin Approach this probability becomes higher than 90%.

Table 8.4. Percentage of Poor by Gender of Head of Household

	Male Head	Female Head	Male or Female Head
Share in Total Number of Households	69.3	30.7	100.0
TFA Approach	21.1	30.0	23.8
TFR Approach	23.7	32.7	26.4
VW Approach	27.1	28.6	27.5
Total Number of Households	141501	62597	204098

Table 8.5. Percentage of Poor by Household Size

	1	2	3	4	5	6	7	8	9	10 +	Total
Share in Total Number of Households	19.5	23.4	15.6	17.4	12.9	6.1	2.5	1.3	0.7	0.5	100.0
TFA Approach	44.5	24.5	18.4	13.1	12.2	17.2	26.4	34.8	40.9	44.6	23.8
TFR Approach	45.9	26.8	21.2	15.7	14.8	20.9	31.5	41.2	47.6	52.4	26.4
VW Approach	33.2	23.0	21.4	20.9	24.2	35.1	54.8	71.3	84.9	94.3	27.5
Total Number of Households	39816	47827	31900	35432	26237	12500	5196	2627	1507	1056	204098

The Impact of the age of the head of the household:

In Table 8.6 we look at the effect of the age of the head of the household on the probability of a household being poor. It appears that, whatever approach one selects, this probability is highest for households whose head is less than 30 years old. According to the TFA and TFR approaches this probability is lowest when the head of the household is 30 to 59 years old while the Vero-Werquin approach indicates that it is lowest for households headed by an individual 60 to 69 years old.

The role played by the marital status of the head of the household: This impact is analyzed in Table 8.7. It appears that according to each of the three fuzzy approaches the probability for a household to be poor is lowest when the head of the household is married. Note that Table 8.7 also indicates that according to any of the three fuzzy approaches adopted poverty is highest when the head of the household is single.

Table 8.6. Percentage of Poor by Age of Head of Household

	less than 30	30-59	60-69	70 at least	Total
Share in Total Number of Households	13.2	56.4	13.8	16.6	100.0
TFA Approach	47.1	17.5	20.7	29.5	23.8
TFR Approach	50.1	20.1	23.2	31.9	26.4
VW Approach	44.8	25.3	21.5	26.3	27.5
Total Number of Households	26873	115107	28242	33876	204098

Table 8.7. Percentage of Poor by Marital Status of Head of Household

	Married	Divorced	Widowed	Single	Total
Share in Total Number of Households	70.1	7.5	13.7	8.7	100.0
TFA Approach	17.5	33.4	31.8	54.3	23.8
TFR Approach	20.2	36.9	34.1	55.7	26.4
VW Approach	25.2	31.0	26.1	45.4	27.5
Total	143010	15368	28011	17709	204098

The Impact of the Year of Immigration: In Table 8.8 we examine the role played by the fact that the head of the household is a new immigrant or not as well as the impact of the year in which he/she immigrated to Israel. The data of Table 8.8 clearly indicate that the more recently the head of the household immigrated, the more likely it is that the household will be considered as poor, this conclusion being true for any of the three fuzzy approaches. Note in particular that among the households whose head immigrated in 1995, the year the Census was taken, more than 60% of them are poor, whatever the fuzzy approach selected.

The Influence of Education Level

Table 8.9 gives the impact of the level of education of the head of the household on the probability of the household being considered poor. Here the results are also straightforward. The higher the level of education of the head of the household, the lower the probability of the household being considered poor, this being true according to any of the three approaches (TFA, TFR and VW) adopted. One may note in particular that among households whose head had no schooling, poverty is more than twice as high as among those whose head had 13 or more years of education. The level of schooling of the head of the household seems thus to be a very good indicator of the wealth of the household or at least of its ability not to be counted among the poor households.

Table 8.8. Percentage of Poor by Year of Immigration of Head of Household

	Before 1990	1990	1991	1992	1993	1994	1995	Total
Share	86.3	4.2	3.5	1.7	1.6	1.7	1.1	100.0
TFA Approach	21.7	25.8	31.4	38.5	45.4	51.6	66.4	23.8
TFR Approach	24.1	29.5	34.3	42.7	49.5	56.5	70.3	26.4
VW Approach	26.0	25.8	31.6	38.8	43.7	50.5	64.5	27.5
Total Number of Households	176043	8473	7137	3437	3285	3460	2263	204098

Before 1990 the data also include the natives.

Table 8.9. Percentage of Poor by Years of Schooling of Head of Household

	0	1-8	9-12	13 or more	Total
Share in total Number of Households	6.4	19.9	41.4	32.3	100.0
TFA Approach	45.8	27.2	21.7	20.1	23.8
TFR Approach	48.9	30.6	24.6	21.8	26.4
VW Approach	47.3	33.9	26.5	20.9	27.5
Total Number of Households	13145	40564	84483	65906	204098

The Impact of Participation in the Labor Force:

In Table 8.10 we indicate the relationship that exists between the probability of a household being considered poor and the number of months its head worked during the last twelve months. It appears, at least according to the Totally Fuzzy Absolute and the Totally Fuzzy and Relative Approach, that the greater the number of months the head of the household worked during the last twelve months, the lower the probability of the household being considered poor. Curiously the Vero-Werquin approach, though indicating that poverty is lowest when the head of the household worked 9 to 12 months during the past twelve months, shows that it is highest when he/she worked 5 to 8 months rather than less than 5 months.

The Impact of the Status at Work:

Table 8.11 indicates the impact of the status at work of the head of the household on the probability of the household being considered poor. Not surprisingly it appears that this probability is highest when the head of the household does not work. The data in Table 8.11 also indicate that this probability is lowest when the head of the household is self-employed. Note that the Totally Fuzzy and Absolute as well as the Totally Fuzzy and Relative approach indicate that among self-employed heads of households the probability of the household being poor is three times lower than that which is observed when the head of the household does not work.

Table 8.10. Percentage of Poor by number of months worked by the head of the household during the last 12 months

	4 or less	5-8	9-12	Total
Share in total number of households	40.1	4.3	55.6	100.0
TFA Approach	33.1	32.0	16.6	23.8
TFR Approach	36.3	35.2	18.6	26.4
VW Approach	33.9	37.7	22.1	27.5
Total Number of Households	81905	8789	113404	204098

Table 8.11. Percentage of Poor by Status at Work of the Head of the Household

	Not working	Salaried	Self-employed	Other	Total
Share in total number of households	37.2	51.7	9.9	1.2	100.0
TFA Approach	32.7	19.9	10.9	26.6	23.8
TFR Approach	36.0	22.2	12.3	28.9	26.4
VW Approach	33.4	25.1	18.3	27.5	27.5
Total Number of Households	76019	105461	20252	2366	204098

The Role of the Place of Residence

In Table 8.12 we indicate the impact of the place of residence on the probability of a household being poor. Whatever the fuzzy approach adopted it appears that poverty is highest when the household is located in Jerusalem and lowest in cities with 100,000 to 200,000 inhabitants.

Table 8.12. Percentage of Poor by Place of Residence

	Jerusalem	Tel-Aviv	Haifa	Cities with 100,000 to 200,000 inhabitants	Cities with 20,000 to 100,000 inhabitants	Municipalities with 2 to 20,000 inhabitants	Other	Total
Share in total number of households	8.9	9.7	6.5	26.0	29.4	14.3	5.3	100.0
TFA Approach	34.4	28.3	23.8	17.8	21.2	27.9	31.8	23.8
TFR Approach	37.6	31.1	26.7	20.1	23.8	30.4	33.6	26.4
VW Approach	41.7	35.9	29.9	20.0	22.7	32.8	34.8	27.5
Total Number of Households	18077	19882	13263	53022	59917	29215	10722	204098

Table 8.13. Percentage of Poor by Religion of the Head of the Household

	Jewish	Moslem	Christian	Druze	Other	Total
Share in total number of households	85.1	10.7	2.0	1.0	1.2	100.0
TFA Approach	20.2	48.4	28.3	30.7	48.1	23.8
TFR Approach	22.7	51.7	32.3	32.8	53.0	26.4
VW Approach	22.5	60.5	43.3	42.9	52.5	27.5
Total Number of Households	173668	21863	4013	2091	2463	204098

The Impact of the Religion of the Head of the Household:
This impact is analyzed in Table 8.13. It appears that all the three fuzzy approaches indicate that poverty is lowest when the head of the household is Jewish. The three approaches show also that the highest poverty levels are observed either when the head of the household is Muslim or when he is neither Jewish, Muslim, Christian nor Druze. This case refers probably

to the numerous households headed by new immigrants from the former Soviet Union who are not Jewish. It is clear that what is likely to play a role in this case is the fact that the head of the household is a new immigrant rather than his religion. This is why a regression type of analysis should be conducted in order to determine the specific impact, ceteris paribus, of each factor taken into account. This is precisely the goal of the next section.

Results of the Logit Regression Analysis

The technique

It is well known that when the dependent variable is limited to the interval [0,1], traditional linear regression should not be used. The analysis should rather be based, for example, on the use of the logistic function where the probability that a dependent variable is equal to 1, given a set of exogenous variables X, would be expressed as: $P(Y=1/X) = 1/ (1 + e^{-Z})$, where $Z= \beta X$.

It is easy to see that when X varies form $-\infty$ to $+\infty$, P will vary from zero to one. Note also that $P/(1-P) = e^{Z}$.

In the following analysis different types of dependent variables have been successively introduced: the probability of being poor according to the TFA fuzzy approach, the TFR fuzzy approach and the Vero-Werquin fuzzy approach.

The results of the estimation procedure:

The exogenous variables that have been taken into account are essentially those whose link with the probability of being poor has been examined in the cross-tables presented in the previous section. More precisely we have introduced, as exogenous variables, the size of the household and its square, the age of the head of the household and its square, the number of years of schooling, the gender, the religion (three dummy variables), the marital status (three dummy variables) and the status at work (working or not) of the head of the household, the area of residence of the household (three dummy variables corresponding to the three big cities) and a variable indicating whether the head of the household immigrated to Israel after 1989. We have also introduced four interaction variables, three between the gender and the marital status of the head of the household and one between the gender and his/her working status.

Table 8.14. Results of the Logit regressions on the basis of a random sample of 30,000 observations

	TFA Approach		TFR Approach		VW Approach	
Variable	coefficient	t- value	coefficient	t value	coefficient	t value
Intercept	5.3309	27.20	5.36123	27.93	3.99908	21.1
Number of years of schooling	-0.0748	-19.20	-.07544	-20.1	-0.08467	-22.7
Household size	-0.6738	-20.08	-.56560	-17.2	-0.29304	-7.85
Square of household size	0.05664	16.59	0.05256	15.50	0.06149	14.7
Age of head of household	-0.14244	-25.82	-.14460	-26.8	-0.12735	-23.7
Square of age of head of household	0.00113	21.30	0.00114	2.12	0.00103	9.8
Head of household is male	0.01217	0.12	-.04045	-0.41	0.15800	1.53
Head of household is Jewish	-.065463	-5.43	-0.61688	-5.17	-0.87776	-7.47
Head of household is Muslim	0.94330	7.25	0.78976	6.15	0.31908	2.53
Head of household is Christian	0.40318	2.57	0.36943	2.40	0.15565	1.05
Head of household is Druze	0.42428	2.29	0.25737	1.42	-0.04773	-0.27
Head of household immigrated after 1989	1.25738	28.69	1.24475	29.44	0.98562	23.6
Head of household is married	-0.35169	-4.33	-.38103	-4.91	-0.08017	-1.03
Head of household is divorced or separated	0.38716	4.80	0.41590	5.32	0.49995	6.18
Head of household is single	0.75819	8.32	0.69963	7.82	0.83457	9.22
Household lives in Jerusalem	0.48399	9.25	0.47334	9.32	0.65894	13.24
Household lives in Tel-Aviv	0.16328	3.13	0.22189	4.42	0.84245	17.84
Household lives in Haifa	0.29262	4.76	0.28475	4.78	0.71793	12.66
Head of household is working	-0.75695	-11.26	-0.79147	-12.3	-0.57461	-9.01
Head of household is male and married	0.17884	1.47	0.23482	1.99	-0.02939	-0.24
Head of household is male and divorced	0.52673	3.46	0.51711	3.45	0.38933	2.52
Head of household is male and single	0.00013	0.00	0.09076	0.65	-0.12128	-0.85
Head of household is male and working	0.04260	0.55	-0.00201	- 0.03	- 0.11755	-1.61
Pseudo R^2	0.20246		0.18787		0.16737	

The results of this logit regression are given in Table 8.14 and are based on a random sample of 30,000 observations[1].

To have an idea of the goodness of fit of the logit regressions a criterion was used that is similar to the R-square used in linear regressions. The idea is to compute the maximal value of the log-likelihood (ln L) and compare it with the log likelihood obtained when only a constant term is introduced ($\ln L_0$).

The likelihood ratio LRI is then defined as:

$$LRI = 1 - (\ln L / \ln L_0) \tag{8.4}$$

The bounds of this measure are 0 and 1 (Greene 1993, pp. 651-653).

It appears that almost all of the explanatory variables that have been introduced have a significant impact which corresponds in fact to the results that have been observed earlier when we compared only the binary link that exists between the probability of a household being considered poor and a given explanatory variable. Thus households whose head has a higher educational level have, ceteris paribus, a lower probability of being poor. This probability decreases and then increases again with the size of the household as well as with the age of the head of the household. Other constant factors also observed is the probability that a household is considered as poor is highest among heads of household that are Muslims and lowest among those who are Jewish; this probability is lowest when the head of the household is married and highest when he/she is single; it is higher when he/she is a new immigrant, is highest when he/she lives in Jerusalem and lowest when he/she lives outside the three main cities; finally ceteris paribus the probability that a household is considered poor is higher when its head does not work. The gender of the head of the household was found to have no significant impact on the probability of being poor. Most of the interactions were not significant. Note however that heads of households who are male and divorced have a higher probability of being poor, whatever the fuzzy approach one adopts.

[1] It was necessary to work only with a sample and not with the population because otherwise we would not have been able to implement the Shapley procedure that will be described later on.

8.3.5 The Shapley Approach to Index Decomposition and its Implications for Multidimensional Poverty Analysis

The Concept of Shapley Decomposition

Let an index I be a function of n variables and let I_{TOT} be the value of I when all the n variables are used to compute I. I could for example be the R-square of a regression using n explanatory variables, any inequality index depending on n income sources or on n population subgroups.

Let now $I_{/k}^{k}$ (i) be the value of the index I when k variables have been dropped so that there are only (n-k) explanatory variables and k is also the rank of variable i among the n possible ranks that variable i may have in the n! sequences corresponding to the n! possible ways of ordering n numbers. We will call $I_{/(k-1)}^{k}$ (i) the value of the index when only (k-1) variables have been dropped and k is the rank of the variable (i).

Thus $I_{/1}^{1}$ (i) gives the value of the index I when this variable is the first one to be dropped. Obviously there are (n-1)! possibilities corresponding to such a case. $I_{/0}^{1}$ (i) gives then the value of the index I, when the variable i has the first rank and no variable has been dropped. This is clearly the case when all the variables are included in the computation of the index I.

Similarly $I_{/2}^{2}$ (i) corresponds to the (n-1)! cases where the variable i is the second one to be dropped and two variables as a whole have been dropped. Clearly $I_{/2}^{2}$ (i) can also take (n-1)! possible values. $I_{/1}^{2}$ (i) gives then the value of the index I when only one variable has been dropped and the variable i has the second rank. Here also there are (n-1)! possible cases.

Obviously $I_{/(n-1)}^{n}$ (i) corresponds to the (n-1)! cases where the variable i is dropped last and is the only one to be taken into account. If I is an inequality index, it will evidently be equal to zero in such a case. But if it is for example the R-square of a regression it would give us the R-square when there is only one explanatory variable, the variable i. Obviously $I_{/n}^{n}$ (i) gives the value of the index I when variable i has rank n and n variables have been dropped, a case where I will always be equal to zero by definition since no variable is left. Let us now compute the contribution $C_j(i)$ of variable i to the index I, assuming this variable i is dropped when it has rank j. Using the previous notations $C_j(i)$ is defined as:

$$C_j(i) = (1/n!) \sum_{h=1 \text{ to } (n-1)!} [I_{/(j-1)}^{j}(i) - I_{/j}^{j}(i)]^{h} \qquad (8.5)$$

where the superscript h refers to one of the (n-1)! cases where the variable i has rank j.

The overall contribution of variable i to the index I may then be defined as:

the household is married and highest when he/she is single; it is higher when he/she is a new immigrant, is highest when he/she lives in Jerusalem and lowest when he/she lives outside the three main cities; finally ceteris paribus the probability that a household is considered as poor is higher when its head does not work. The gender of the head of the household was found to have no significant impact on the probability of being poor.

The Shapley decomposition gave some additional information because it showed that the three categories of variables that had the greatest impact, whatever the fuzzy approach adopted, were the age of the head of the household, his/her religion and the size of the household. Other variables that played an important role, though their relative importance was smaller, were the year of immigration, the educational level and the marital and working status of the head of the household. Finally the impact of the place of residence and of the gender of the head of the household was small.

Finally this study also revealed that 18.5% of all the households were poor according to the three approaches. 25.6% according to at least two approaches and 33.7% according to at least one approach. Thus even though there was a good amount of overlapping between the three distributions, they were far from identifying the same households as poor.

To find the Shapley contribution of each factor a specific combination is run twice, with the factor and without it. Therefore we run 13,824=256x2x9x3 logit regressions (9 are the factors and 3 is TFA, TFR and VW).

Bibliography

Cerioli A, Zani S (1990) A Fuzzy Approach to the Measurement of Poverty. In: Dagum C, Zenga M (eds) Income and Wealth Distribution, Inequality and Poverty, Studies in Contemporary Economics. Springer, Berlin Heidelberg, New York, pp 272-284

Chantreuil F, Trannoy A (1999) Inequality Decomposition Values: The Trade-Off Between Marginality and Consistency. THEMA Discussion Paper, Université de Cergy-Pontoise

Cheli B, Ghellini G, Lemmi A, Pannuzi N (1994) Measuring Poverty in the Countries in Transition via TFR Method: The Case of Poland In 1990-1991. Statistics in Transition 1:585-636

Cheli B, Lemmi A (1995) A "Totally" Fuzzy and Relative Approach to the Multidimensional Analysis of Poverty. Economic Notes 24:115–134

Ragin CC (2000) Fuzzy-Set Social Science. The University of Chicago Press, Chicago

Sastre M, Trannoy A (2002) Shapley Inequality Decomposition by Factor Components: Some Methodological Issues. Journal of Economics Supplement 9:51-89

Shorrocks AF (1999) Decomposition Procedures for Distributional Analysis: A Unified Framework Based on the Shapley Value. Mimeo, University of Essex

Silber J (1999) Handbook on Income Inequality Measurement. Kluwer Academic Publishers, Dordrecht and Boston

Sorin M (1999) Multidimensional Poverty Analysis. Using the Fuzzy Sets Theory. MA Thesis, Department of Economics, Bar-Ilan University, Ramat-Gan, Israel

Vero J (2002) Mesurer la pauvreté à partir des concepts de biens premiers, de réalisations primaires et de capabilités de base. Thèse de doctorat, Ecole des Hautes Etudes en Sciences Sociales, Groupement de Recherche en Economie Quantitative d'Aix-Marseille (GREQAM)

Vero J, Werquin P (1997) Reexamining the Measurement of Poverty: How Do Young People in the Stage of Being Integrated in the Labor Force Manage (in French). Economie et Statistique 8-10:143-156

Zadeh LA (1965) Fuzzy Sets. Information and Control 8:338-353

Appendix: List of Variables available in the 1995 Israeli Census

Number of rooms:

1: 1 room
2: 1.5 rooms
3: 2 rooms
4: 2.5 rooms
5: 3 rooms
6: 3.5 rooms
7: 4 rooms
8: 4.5 rooms
9: 5 rooms
10: 5.5 rooms
11: 6 or more rooms

Year of construction of dwelling:

1: Before 1947
2: 1948-1954
3: 1955-1964
4: 1965-1974

5: 1975-1984
6: 1985-1989
7: 1990
8: 1991
9: 1992
10: 1993
11: 1994
12: 1995

Ownership of dwelling:

1: family owned
2: rented

Bath/Shower:

1: Bath (with/without shower)
2: Shower only
3: No bath or shower

Telephone

1: Yes
2: No

Television

1: Yes
2: No

Videotape

1: Yes
2: No

Washing Machine

1: Yes
2: No

Microwave Oven

1: Yes
2: No

Dishwasher

1: Yes
2: No

Computer

1: Yes
2: No

Air-Conditioning

1: Yes
2: No

Solar Heating System

1: Yes
2: No

Dryer

1: Yes
2: No

Cars available

1: No car
2: 1 car
3: 2 cars
4: 3 cars or more

9 Multidimensional Fuzzy Set Approach Poverty Estimates in Romania

Maria Molnar, Filofteia Panduru, Andreea Vasile, Viorica Duma

Maria Molnar
Institute of National Economy, Romanian Academy
Filofteia Panduru, Andreea Vasile, Viorica Duma
National Institute of Statistics

9.1 Introduction

Poverty is one of the most serious social problems Romania is dealing with. During the transition from the planned to market economy, as a result of economic decline, the living standard has fallen, poverty has increased and population poverty and feeling of deprivation has marked all the transition period. In order to avoid a poverty explosion, economic reforms have to some extent slowed down, with heavy impact on national economy performance. A social protection system had to deal with increasing tasks whilst resources have decreased.

In this context, the measurement and analysis of poverty became a subject of great concern for researchers, statisticians, and also for social as well as economic policy makers. Dealing with poverty measurement grew on the occasion of the development of national strategies for poverty alleviation – National Strategy for Poverty Preventing and Alleviation (1998) and National Anti-Poverty and Social Inclusion Plan (2002). The activities related to measuring and fighting poverty are co-ordinated by the Anti-Poverty and Social Inclusion Commission (CASPIS), set up in 2001 by the Romanian Government. The research projects on poverty had financial and scientific support from UNDP, European Commission (PHARE Programme)[1], World Bank, DFID, etc.

In Romania, in the evaluation of poverty dimensions several methods are used. CASPIS estimates poverty using a method developed and usually applied by World Bank experts. This method is based on the evaluation of

---

[1] As part of the Phare programme, the Romanian National Institute for Statistics experts involved in poverty evaluation received technical assistance from the statistical institute of France (INSEE) and Italy (INSTAT), as well as from Siena University experts.

a minimum food basket and on the estimate of two non-food expenditure minimum amounts, which represent poverty and severe poverty thresholds.

The National Institute for Statistics evaluates poverty by the standard relative method used to estimate the structural and Laeken poverty indicators. Poverty has also been estimated and analysed using the multidimensional fuzzy set approach (INS et al. 1998; UNDP 1999; Panduru, Molnar and Gheorghe 2000).

This Chapter begins with a presentation of the demographic and economic context of poverty analysis in Romania and of the main results of traditional monetary poverty evaluation. Results of multidimensional fuzzy set approach to poverty measurement in Romania are examined in the second section of the Chapter.

9.2 Socio-economic and demographic context

At the beginning of 2004, the Romanian population was 21.7 million. During the last decade, the population decreased year after year, thus in 2004 the number of people was by 1.1 million smaller compared with 1992. Starting with 1992, the negative natural increase was the major contribution to the decline of population, due to the large fall in the birth rate.

Demographic ageing also affects the Romanian population. The average age rose from 35.1 years in 1992 to 38.2 years in 2004. The young population (up to 15 years old) decreased by over 1.6 million, its share in the total population fell from 22.7 percent in 1992 to 16.4 percent in 2004. Instead, the population of 65 year olds increased by 625 000, its share rose from 11.0 percent to 14.4 percent. Thus, if in 1992 for 100 persons up to15 there were 48 persons aged 65 years and over, in 2004 this rate had doubled.

As far as the household size is concerned, the data of Households Budget Survey (2003) reveal the great number of single persons (22.8 percent of total households), especially old women, and the predominance of the households with two, three and four persons (26.3 percent, 20.8 percent and 17.9 percent, respectively). The share of households with five or with six and more persons is lower (6.9 percent and 5.3 percent, respectively). More than two thirds of households have no dependent children (69.3 percent), 19.1 percent have one child, 8.9 percent have two children and 2.7 percent have three or more children. A quarter of total households (26.5 percent) are headed by women, a lot of them single.

Almost half of the households have as reference person a pensioner (42.0 percent), a third – an employee/wage earner (32.6 percent) and one in ten – a farmer (10.0 percent). Households with an unemployed reference person were 4.8 percent, those headed by employers or self-employed in non-agricultural activities – 0.6 and 3.1 percent, respectively.

The present socio-economic structure of Romania is mainly determined by the transition from the centralized planned to the market economy. In the period of transition the national output fell in almost every year. The gross domestic product decreased strongly during the early stage of transition, recording, in 1992, 79.4 percent of the level in 1990. After a period of four years of increase (1993-1996), to 93.4 percent of the 1990 year level, the gross domestic product fell again (by 6.9 percent in 1997, by 7.3 percent in 1998 and by 1.2 percent in 1999).

The fall in production has been accompanied by a severe contraction of employment. The civilian labour force employment decreased from 10.8 million in 1990 to 8.6 million in 2000 (38.5 percent of total population) and to 8.3 million persons in 2003. In 2003 the number of wage earners was about half that in 1990 (4.7 against 8.1 million).

Thus, unemployment exploded and affected over a million persons (more than 10 percent of the active population) in the years 1993-1994 and 1998-2000. The unemployment rate was more than 8 percent throughout 1992-2000, with the exception of the year 1996 (6.6 percent). The number of pensioners also increased (from 3.6 million in 1990 to 5.9 million in 2000). Retirement was one of the ways of reducing tension on the labour market, through early retirement and easing eligibility for invalidity pension. As employment, especially the number of employees, decreased, the number and the proportion of population with low incomes - earned in agricultural self-employment or received as social benefits (pensions, unemployment benefit or social assistance benefits) - increased and the poverty risk extended.

Throughout the transition, the living standard was also lowered by the high (two or three digits) inflation that determined the decreasing of purchasing power of all incomes. The fall of real incomes was stronger over 1991-1994, when the consumption prices rose on average more than 70 times, thus the real earning and real pension fell by 40.6 and 44.7 percent, respectively. In 1995 and 1996, as the GDP increased and inflation decreased (as a result of the economy restructuring slow down and of the fact that 1996 was an electoral year), the real earnings rose by 11.9 and 9.4 percent, and the real pension by 10.7 and 2.6 percent, respectively. Restarting the economic restructuring process in 1997, together with a new inflation wave, brought about o new decrease of the real earnings (by 22.6 percent) and of the real pensions (by 20.7 percent). In 1999, the real earnings were 56.0 percent of the 1990 year level (84.3 percent of that in 1995), and real state social insurance pension decreased up to 2000 to 43.9 percent of the 1990 year level (71.7 percent of that in 1995).

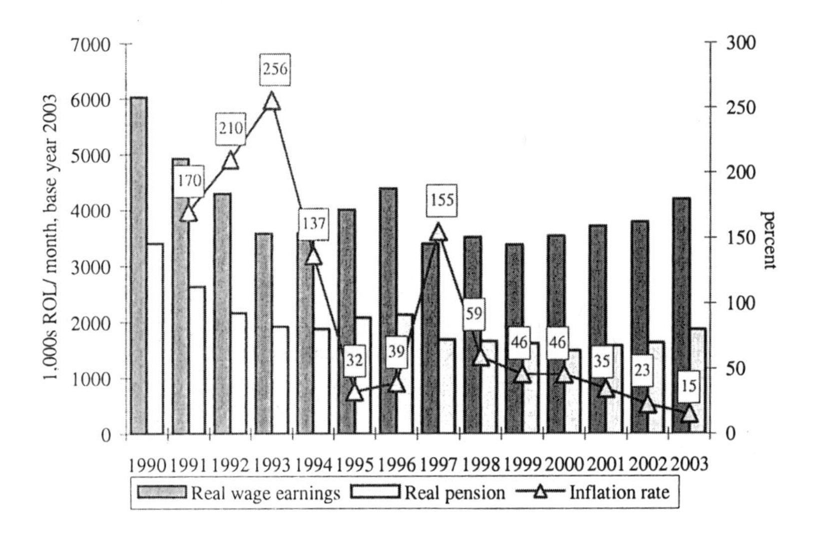

Fig. 9.1. Real earnings and pensions, 1999-2003. Source: Statistical Yearbook

According to a Household Budget Survey, the households real disposable incomes decreased in the year 2000 by 27.3 percent compared to the year 1995. The income decrease was experienced by a large part of the population, by all categories, thus the number of those persons that couldn't cover the expenditures necessary to ensure a minimum living standard increased, and the poverty risk was amplified.

Economic growth started again with the year 2000: the gross domestic product increased by 2.1 percent in 2000, by 5.7 percent in 2001, by 5.1 percent in 2002 and by 5.2 percent in 2003[2]. Civilian employment continued to decrease, and the number of pensioners rose in 2001 and 2002, recording a slight decrease in 2003. Unemployment decreased considerably: the number of unemployed diminished from over a million in 2000 to 660,000 in 2003 and the unemployment rate reduced from 10.5 to 7.4 percent. Beginning with the year 2002 the number of employees increased slightly.

Economic growth and the slow down of inflation (from a rate of 45.7 percent in 2000 to 15.3 percent in 2003) have been accompanied by income growth: earnings rose between 2000 and 2003 by 18.7 percent, pensions by 17.3 percent, and household disposable incomes by 13.3 percent. Thus, in 2003 the earnings were 69.5 percent, average pension 51.4 percent and disposable income 82.4 percent of the levels recorded in 1995

[2] In the year 2003, the gross domestic product per inhabitant was 7544 $ PPP.

(86.9 percent for households living in urban areas and 76.2 percent for those living in rural areas).

In 2003, the total gross incomes of the households were over 7.8 million ROL monthly per household, and disposable incomes were 6.1 million ROL per households (2.2 million ROL per person).

Information collected by the Household budget survey in 2003 reveal relatively large differences between household incomes grouped by occupational status of the household head. Estimated as average per person, employers and employees household's income exceed the average by 76 and 23 percent, respectively, incomes of pensioners' households are close to the average income (-3 percent), and incomes of farmer, unemployed and self-employed households are lower than the average values (by 37, 34 and 23 percent, respectively). Disposable incomes of households in urban areas are about 50 percent higher than those in rural areas.

The inequality of disposable income per adult equivalent is relatively small. In 2003, the ratio between the "richest" and the "poorest" quintile was 4.6:1 and the Gini coefficient was 0.296.

9.3 Monetary dimension of poverty

9.3.1 National method

The method of poverty measurement, used by CASPIS, defines the poverty and severe poverty status by two thresholds established on the basis of a minimum food basket and of the food consumption share in the total consumption expenditures. The quantities in the minimum food basket are the same over all the period of poverty evaluation (1995-2003), giving the poverty thresholds the "absolute" character and allowing comparability of the poverty indicators in time.

The minimum basket contains 9 groups of food products (96 items). The quantities for each item are those preferred by households from the bottom of distribution by the consumption expenditures (the second and the third decile), scaled up proportionally in the way to ensure a daily 2550 calorie intake per adult. The food basket is priced at median "unit value" of products purchased by the same group of households.

The non-food component of the poverty line (which is added to the cost of the minimum food basket) is established at the amount spent on non-food and services by households whose expenditure on food consumption equals the cost of the minimum food basket. The non-food component of the severe poverty threshold is established at a lower level, equal to the non-food expenditure of households whose total consumption expenditure matches the minimum food basket in value.

The consumption expenditures per adult equivalent are used as an indicator of households' welfare. Consumption expenditure includes the current expenditures on food, non-food and services, the value of food consumption out of own production, and the use value of some durables.

To ensure comparability across time, the consumption expenditures are adjusted for inflation, by components, using consumer price indices for food, non-food and services. The expenditures of rural households are accordingly adjusted for rural-urban differences in living cost. Moreover, a seasonality index is used to smooth expenditure in order to diminish the impact of seasonality in consumption on the comparability of households recorded in different months[3].

The equivalence scale is calculated according to formula $AE = (A + 0.5C)^{0.9}$, where A represents the number of adults in the household, C – the number of children. Parameters – $\alpha = 0.5$ and $\theta = 0.9$ – were estimated on the basis of a relation between the economies of scale parameter, the proportion of private goods in total household consumption and the household size[4]. A regression model on the dependence of adult goods to the total consumption, the household size and the share of adults, children and older persons has also been used[5].

According to this method, in 2003 the poverty rate (estimated for persons) was 25.1 percent and the severe poverty rate was 8.6 percent. The estimation results show a dramatic extending of the poverty incidence in 1997, continuing from 1998 to 2000, and a fair decreasing, by about a third, after 2000. Thus, the proportion of population under the poverty threshold is very sensitive to economic performance, to the decline or to the growth of the GDP, and to the increase or decrease of inflation. Such changes in the poverty rate derive also from the relatively high frequency of households near (above or under) the poverty line. As the mean poverty gap is relatively low (between 24 and 27 percent throughout 1995-2003), many households are likely to cross the poverty line.

---

[3] All these adjustments are needed because the Household Budget Survey is carried out on a sample divided in monthly waves: each household is interviewed every month.

[4] The relation was proposed by Lanjouw and Ravallion (1994).

[5] The equivalence coefficients of this scale are generally close to those contained in another scale proposed in a study on equivalence scale in Romania (Betti, Molnar and Panduru 2003).

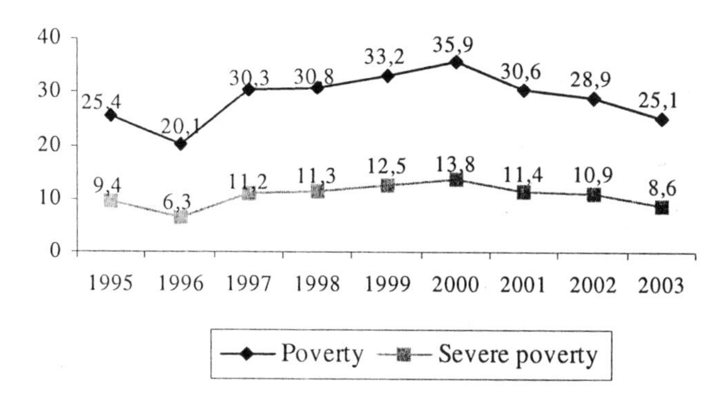

Fig. 9.2. Headcount poverty rate (percent). Source: CASPIS 2004

The poverty incidence is higher in rural areas: in 2003, the poverty rate of the rural population was 38.0 percent, and the severe poverty rate – 13.9 percent, compared with 13.8 and 3.8 percent, respectively, in urban areas. Poverty is most frequent in the regions in which the rural population surpasses the urban population (35.4 percent in the North-East and 32.1 percent in the South-West) and exceeds the average in the South (29.9 percent) and also in the South-East (29.2 percent). In the Centre, West and North-West regions the poverty rate was lower (20.3, 18.1 and 17.7 percent, respectively), and in Bucharest – 8.1 percent.

Households with five or more persons and especially those with three and more dependent children record the highest poverty rate (45.1 and 56.8 percent) at a considerable distance to those recorded by the households with 1-4 persons (17.8, 17.5, 16.0 and 21.1 percent) and to those without children or with one or two children (21.4, 23.1 and 28.7 percent).

Young persons 15-24 years old and children (0-14 years) are more frequently exposed to the poverty risk (31.9 and 29.9 percent) than persons aged 25-64 (21.6 percent) or older (24.9 percent).

Regarding the poverty incidence by occupational status of the household head, it is obvious that the persons living in farmers' households are more likely to be poorer than those living in unemployed' or pensioners' households (48.2 percent against 34.4 and 21.9 percent, respectively). Poverty rates also exceed the average among households headed by self-employed persons in non-agricultural activities (30.2 percent). Poverty is less widespread among employees' households, due to the fact that, even if the general level of salaries is not high, salary incomes are slightly higher than all the other incomes (with the exception of some incomes from independent activities or from property).

The poverty risk is connected with low level of education, which diminishes the chance of employment or of getting jobs with high income and, on the other hand, generate a demographic behaviour leading to numerous families with a lot of children. In 2003 the poverty rate was 55.7 percent among households where the reference person didn't even graduate from primary school, 37.7 and 29.0 percent among those headed by persons who graduated only from primary or lower secondary school. The upper secondary education level of the reference person corresponds to poverty rates of 19.3 and 10.3 percent, and the higher level of education – with 1.5 percent.

9.3.2 Relative method

The National Institute of Statistics program on the estimation of structural and social inclusion indicators contains the evaluation of poverty using the standard relative method, according to the European Commission and Eurostat experts recommendations. This method estimates the poverty indicators using disposable household incomes (including consumption from own resources), adjusted by the modified OECD equivalence scale, and the poverty risk line established at 60 percent of median incomes.

According to this method[6], the 2003 year poverty risk rate has been estimated at 17.3 percent[7]. The proportion of population under the 40% cut-off was 5.0 percent, that of the population with income between 50 and 60% cut-off – 7.1 percent, and that of the population living in households with income between 60 and 70 % cut-off – 7.9 percent. The relative median at the risk-of-poverty gap was estimated at 21.5 percent.

The poverty risk rate was 17.1 percent in 2000, 17.0 percent in 2001 and 18.1 percent in 2002. Estimated by the 2000 year threshold, it emphasizes the decrease of the poverty risk incidence: from 17.1 percent in 2000 to 15.5 percent in 2001, 15.1 percent in 2002 and 11.6 percent in 2003.

In 2003, the poverty risk rate was higher for people aged under 16 (22.4 percent), for those aged 16-24 (19.6 percent), and for those aged 65 and over (19.5 percent), especially for the women in this age group (23.6 percent). Poverty risk is more frequent among the self-employed (33.8 per-

[6] Poverty rate, estimated using consumption expenditures, adjusted by CASPIS scale, and the 60% threshold, is 15.3 percent, and that estimated using expenditures, adjusted by the scale proposed by Betti, Molnar and Panduru (2003), is 15.6 percent.

[7] The weight of population living in households whose disposable income, estimated without consumption from own resources, is less than 60% of this income median was 25.4 percent in 2003.

cent) and among the unemployed (30.6 percent), and is less frequent among pensioners (14.7 percent) and employees (3.3 percent).

One out of four single persons (25.1 percent) has an income lower than the poverty risk threshold, the rate being higher for single women (28.4 percent) and for single older people (31.4 percent). The high poverty risk affects the households with two adults and three or more dependent children (42.3 percent); also the households with one adult and one or more dependent child (20.8 percent) and those with three or more adults and one or more dependent child (24.4 percent). The lowest poverty risk rate is recorded for households containing two adults with one child and households with two adults under 65 years (10.5 percent).

9.4 Multidimensional estimation of poverty

Information collected by household surveys shows the high share of households that are facing deprivation regarding housing conditions and lack of main durable goods. Thus, over 40 percent of households does not have access to running water through the public network, and dwellings of over 60 percent of them are not connected to the public heating network and do not have their own heating central. A quarter of the households live in dwellings made of weak materials (adobe etc), and one of six in overcrowded dwellings (more than two persons/room). The frequency of households without the main durable goods varies between 9 percent regarding the TV set and 80 percent regarding the car. There are deprivations resulting from the long term persistence of low incomes, insufficient for the investments required by house arrangement and buying durable goods, on one hand, and from insufficient development of public utilities, especially in rural areas, on the other. Almost two thirds of households are considering that they cannot face the current expenditures with their income, meaning that they are in difficulty in covering the main needs at expected standards. One of three households delays paying the utilities consumption (bills for heating, water and sewerage, electricity), so the well-being due to access to utilities and endowment with home electric appliances is weakened by the stressing difficulty in paying the bills, which have highly increased in the last years.

Given the availability of such statistical information a multidimensional fuzzy set approach to poverty and living condition estimation has been performed referring to the so called TFR method proposed by Cheli and Lemmi (1995) and described in detail in several Chapters of the present Book.

In the Romanian context the TFR estimates are obtained on the basis of a quantitative variable (consumption expenditures by adult equivalent) and 18 qualitative variables. The qualitative variables refer to:

- the quality of housing conditions: lack of running water, hot running water and central heating (access to public network or private endowment), lack of bathroom and indoor flushing toilet, over-crowding (more than two persons/room) and dwelling built from weak materials;
- lack of durable goods (TV set, radio, car, phone, refrigerator, washing machine, vacuum cleaner, gas cooking stove);
- financial difficulties (impossibility to cover current expenditures by disposable income, delays in paying bills, impossibility to buy sufficient food products).

Multidimensional estimation refers to 2003 and is based on information collected by the Household Budget Survey, which collects information on the well-being monetary indicators (household incomes and consumption expenditures), on the main components of living conditions, and also on financial difficulties.

In view of outlining the profile of poor households, poverty TFR indices (global and partial) were calculated by groups of households: by occupational status, level of education, sex and age of the household head; by household size, number of dependent children and household type; and by area of residence and regions (Table 9.1). A regression model was also estimated, with 27 independent dummy variables referring to occupational status and education level of the household head, household type, area of residence and region in which household is living (Table 9.2). On average, TFR global index is equal to 0.316, and partial indices are 0.350 for living conditions, 0.246 for durable goods endowment, 0.440 for financial difficulties and 0.213 for consumption expenditures.

9.4.1 Poverty and occupational status

The results of poverty estimation by household groups according to occupational status of the household head shows that the poorest households are the agricultural ones, which have a TFR global index of 0.527, 67 percent higher than the mean index. The households headed by a farmer stand out with the worst housing conditions, the weakest endowment with durable goods and with the highest frequency of households with consumption expenditures under the national poverty threshold: partial indices corresponding to the first two groups of deprivation exceed the appropriate average indices by 88 percent and 83 percent, respectively, and the index corresponding to consumption expenditures is more than twofold higher.

Those household living conditions are marked by rural area residence, mostly deprived of the infrastructure necessary for providing public utilities (water and sewerage, natural gas). The multiple deprivations of farmer households is determined mostly by the low level of incomes, especially money incomes, which restrains both the current consumption and especially investment. Considering this aspect, farmers were one of the most disadvantaged categories in the old regime, too: the earnings from working in agricultural cooperatives were generally very low, so they couldn't afford to build dwellings endowed with own equipment for water supply, with central heating, bathroom and indoor flushing toilet, and in many cases not even to use resistant building materials. The precarious living conditions of the farmers has old roots and it is perpetuating as long as agriculture is not a paying activity, being to a great extent a subsistence one. TFR index estimated for financial difficulties is also very high for farmers (0.516), but exceeded by the one estimated for households headed by an unemployed person (0.616) and for households headed by inactive persons, other than pensioners (0.549). Many of them live in urban areas, mainly in blocks of apartments, so utility bills that they must pay are much higher than for farmers. Also, because farmers households cover most of their food consumption from their own production, the pressure of this part of consumption on their budget is also not as high as that on urban households with low incomes.

Another category of employed households for which a global TFR index was estimated above the mean is that of households headed by self-employed persons in non-agricultural activities. The deprivations of households headed by unemployed persons and inactive persons are also larger than those estimated at an average level.

The conditions of lesser poverty are to be found in the households with employers as reference persons, and also in employees' households: the global index estimated for the first category is more than three times lower and that estimated for the latter is slightly over half of the average one.

Thus, unemployed and pensioners households are proven twice poorer than the employees households, and farmers households are proven three times poorer. The outcomes of the multivariate analysis suggest the increase of poverty on average 2.3 times and, respectively, twice for farmers and unemployed households as compared to employees' households, but a low difference between pensioners and employees households (14 percent). The higher level of the global TFR index registered by pensioners households, as compared to employees households, is to a greater extent the outcome of a large number of households made up of elderly single persons (more than a quarter of the total number of pensioners households), and of a combination between pensioner status and rural residence for more than half of pensioners households. The majority of the rural pensioners were farmers, meaning bad housing conditions, a weak endowment with durable goods, and a low level of pension, as the farmers pension is only a fifth of the mean state social insurance pension. Consequently, even if the salary and the social insurance pension level is not high (the proof is the relatively high index for financial difficulties estimated for employees and pensioners households), the presence of the employees in the household and the pensions system are factors that considerably reduce poverty risk, in terms of both monetary and living conditions. The influence of unemployment, social assistance and family benefits is less obvious.

9.4.2 Poverty and education

TFR indices, both global ones and each of the partial ones, show the close relationship between poverty and the low level of education. The greater hardship is found for households whose head has only primary education or has no education. It is 58 percent higher than average and over six times higher than the one faced by households whose reference person has university education[8]. Indices estimated for households headed by persons with lower secondary education at most also exceed the mean.

9.4.3 Poverty and demographic characteristics of households

The extent to which different symptoms of poverty cumulate depends on household size, especially on the number of dependent children. Global TFR index shows the highest poverty degree for households with six or more persons (53 percent higher than the mean one), and for households with four and more children (double as against the mean index).

Households with three members and those with one dependent child, especially households containing two adults and one child are in the best situation. Households without children are, on average, poorer than those with a child, because they are cumulating more deprivations on housing conditions and endowment with durable goods, the differences related to consumption expenditures and financial difficulties being non-significant. Among households without children, those with single persons aged 65 and over (23.0 percent of households without children), and those with two adults, both or at least one aged 65 and over (11.8 and 8.4 percent of

[8] According to the 2003 year Household Budget Survey, 19.8 percent of households were headed by persons with the lowest level of education, of which 16.4 percent were pensioners and 2.2 percent farmers.

households without children) meet more deprivations: global TFR indices corresponding to these categories (0.404, respectively 0.352 and 0.327) are higher than the mean index for all households. On the contrary, the poverty degree of households with two adults under 65 years (23.5 percent) and with three or more adults without dependent children (15.9 percent) is lower than the one of households with a child.

Among households with children, those with two adults and three or more children are in the worst situation: all partial indices for this category are higher than indices estimated for all other household types. Households with three or more adults with children – generally multigenerational households in rural areas, which include old persons without incomes or with low pensions – are in a worse situation than other categories of households with children, both with regards to consumption expenditures and financial difficulties, housing conditions and possession of durable goods[9]. Unexpectedly, the outcomes of the estimation suggest a relatively low poverty degree (under the average) for the households with one adult and children (one parent families, especially), with TFR indices for consumption expenditures and housing conditions lower than the mean, but with financial difficulties bigger than those estimated for the mean. However, multivariate analysis shows that one parent families expect an increase of poverty degree (as against households with two adults and a child) equal to that corresponding to households with three or more adults and children.

Households whose head is old (60 years and over) are poorer than those headed by younger persons, both in monetary terms and as far as housing conditions are concerned; households whose head is aged under 30 years or 41-50 years are in the less unfavorable situation. Household head age is not, however, a factor with significant influence on poverty. As resulted from the estimation of a regression model which includes this variable, an old household head implies the reduction of the poverty degree in relation to that for age groups 30-40 and 41-50. Sex of household head is also not an important factor for the poverty degree. The fact that TFR indices estimated for households headed by women are higher than those estimated for households headed by men is related to the high share of households containing a single women (57.3 percent of households headed by women in 2003), especially older women from rural areas. A regression parameter estimated for the "household headed by a woman" variable is very low, showing an increase of the poverty degree by only 3 percent, on average, as compared to households headed by men, if all other factors are unchanged.

[9] These are 28 percent of households with children.

9.4.4 Territorial distribution of poverty

All TFR indices show the great difference between poverty in urban and rural areas of residence: the global index is 2.6 times higher in rural than in urban areas, indices for consumption expenditures and possession of durables are 2.7 times higher, and the index that measures deprivations related to housing conditions – four times higher; only the index estimated for financial difficulties is almost the same in urban and in rural areas. The rural location is one of the important factors of poverty, determining both monetary and non-monetary dimensions of poverty. The discrepancy has to some extent an historical determination: life conditions were always more difficult, the living standard was much lower and possibilities for human development, mainly training possibilities, were more limited in rural than in urban areas. The discrepancy is maintained now by the low level of efficiency in agriculture and in developing the non-agriculture activities in rural areas, by the occupational structure in rural areas, and by low capability of rural communities to expand the infrastructure of public utilities.

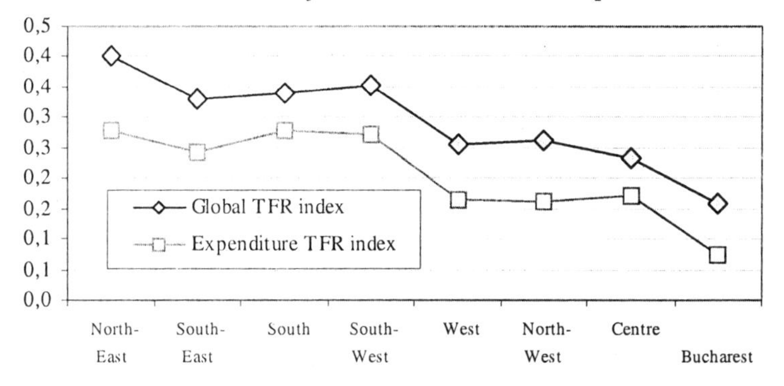

Fig. 9.3. TFR index, by region

The poverty degree also differs according to region. The poorest region is the North-East, with a global index 2.5 times higher than the Bucharest region, which includes Romania's capital. Poverty exceeds the mean in other three regions (the South-West, South and South-East) and is below the mean in the West, North-West and the Centre. The difference between indices estimated for regions from the first and second group is relatively high, as is the difference between indices estimated for Bucharest and all other regions.

The partial index estimated for consumption expenditures in the Bucharest region is lower than half of those estimated for the West, North-West and Centre regions, but the index corresponding to financial difficulties is considerably higher in Bucharest than in the three regions, bringing it close to the mean. Besides the better housing conditions and possession of durables in Bucharest the difficulty is covering the expenses for using them and for covering other basic needs, even if the general level of incomes of Bucharest households is much higher than in all other regions. While the residence in any other region is a factor increasing for monetary poverty in comparison with that registered in Bucharest, from the view of multidimensional estimation of poverty, the picture is different: residence in regions of the North-East, South-East, South and South-West is a factor for increasing poverty when compared to Bucharest, residence in the Centre and North-West is a proven factor for decreasing poverty, and residence in the West does not make a significant difference to poverty when compared to Bucharest.

9.5 Conclusions

Defined by the low level of consumption compared to the minimum necessary for covering consumption needs, poverty affects a quarter of the Romanian population. 17 percent of the population have a disposable income lower than 60% of the median of the population distribution by incomes, considered to the threshold of poverty risk. The frequency of those who cannot cover a minimum of consumption differs from one population category to another, being higher for those for which absence or a weak position on the labour market is added to a large household, especially with many dependent children, and with residence in an economically depressed area.

Beyond the low level of incomes or current consumption, a great share of the population is facing deprivations related to essential elements of living conditions dependent on long term material accumulation and on necessary infrastructure (dwelling and its comfort conditions, household appliances and cultural durable goods, housing environment quality etc). Also a great share of the population have financial difficulties and are unable to cover current expenses with disposable incomes, especially paying the bills for utilities and for buying adequate food products. The deprivations related to living conditions and financial difficulties affect household categories with different intensity and in different ways, cumulating and adding to the low level of incomes and current consumption. Multidimensional analysis of poverty allows a more shading estimation of poverty degree, and outlines the core poverty profile, characterized by more symptoms and dimensions. According to multidimensional fuzzy set approach poverty estimates, households whose head is a farmer, households headed by persons with only primary education or without education, households with six or more persons, especially those with four or more dependent children are the poorest. Poverty is higher in rural areas, and the poorest region is the North-East. Households headed by persons with university education and households headed by employers have the best situation, with the lowest poverty degree; as do households with three persons, especially with two adults and a child, and with two adults under 65 years. Households headed by employees are better off than others too. Bucharest is the least poor region.

Decreasing poverty is obviously related to economic development, to an increasing and improving employment structure, mainly by extending wage employment, which can assure conditions for increasing incomes. Higher incomes for most households means greater possibilities to cover current consumption expenditures, decreasing financial difficulties and increasing possibilities to invest in household comfort and endowment with durable goods. Reducing poverty also imposes the development of the rural economy, modernizing and increasing efficiency in agriculture, development of non-agricultural activities in rural areas and development of the public utilities infrastructure.

Table 9.1. TFR indices in 2003

	[1]	[2]	[3]	[4]	Global
TOTAL	0.350	0.246	0.440	0.213	0.316
Occupational status of the household head					
Employee	0.203	0.129	0.358	0.097	0.178
Employer	0.123	0.067	0.121	0.023	0.089
Self-employed in non-agricult. activities	0.413	0.256	0.444	0.274	0.331
Farmer	0.658	0.451	0.516	0.447	0.527
Unemployed	0.329	0.252	0.616	0.333	0.325
Pensioner	0.388	0.281	0.462	0.225	0.333
Other	0.327	0.338	0.549	0.348	0.359
Educational level of the household head					
Primary	0.592	0.451	0.509	0.380	0.499
Lower secondary	0.472	0.319	0.484	0.284	0.386
Vocational	0.331	0.199	0.464	0.206	0.274
High school	0.212	0.151	0.423	0.118	0.200
Post-high school	0.124	0.099	0.357	0.057	0.134
University	0.061	0.073	0.214	0.023	0.081
Household head sex					
Male	0.349	0.224	0.414	0.207	0.288
Female	0.354	0.306	0.510	0.228	0.339
Age group of the household head					
Under 30 years	0.253	0.235	0.418	0.154	0.256
30 – 40 years	0.326	0.211	0.428	0.183	0.272
41 – 50 years	0.297	0.186	0.441	0.213	0.254
51 – 60 years	0.328	0.215	0.416	0.208	0.275
Over 60 years	0.421	0.316	0.460	0.240	0.362
Household size					
1 person	0.352	0.347	0.480	0.186	0.351
2 persons	0.333	0.239	0.411	0.184	0.286
3 persons	0.232	0.173	0.412	0.186	0.219
4 persons	0.371	0.185	0.428	0.216	0.277
5 persons	0.468	0.242	0.455	0.332	0.349
6 persons and more	0.661	0.350	0.534	0.490	0.485

[1]=Precarious housing conditions; [2]=Absence of endowment with durable goods; [3]=Financial difficulties; [4]=Consumption expenditures

Table 9.1. (cont)

Number of dependent children					
No children	0.329	0.254	0.434	0.200	0.296
1 child	0.312	0.190	0.431	0.204	0.593
2 children	0.485	0.244	0.462	0.253	0.350
3 children	0.652	0.385	0.543	0.436	0.496
4 children and more	0.787	0.521	0.636	0.567	0.626
Household type					
Single person					
under 30 years	0.126	0.235	0.373	0.074	0.202
30-64 years	0.297	0.294	0.460	0.136	0.302
65+ years	0.415	0.394	0.506	0.231	0.401
Two adults without children					
Both of 65 years or over	0.429	0.299	0.424	0.243	0.352
One of 65 years or over	0.395	0.273	0.426	0.211	0.327
Both under 65 years	0.273	0.199	0.396	0.151	0.242
3 or more adults without children	0.296	0.173	0.413	0.231	0.246
One adult with one or more children	0.266	0.267	0.514	0.142	0.286
Two adults with					
1 child	0.217	0.173	0.418	0.135	0.213
2 children	0.445	0.220	0.441	0.187	0.318
3 children and more	0.650	0.421	0.558	0.452	0.515
3 or more adults with children	0.495	0.246	0.469	0.335	0.361
Area of residence					
Urban	0.147	0.139	0.431	0.120	0.174
Rural	0.595	0.375	0.450	0.325	0.453
Region					
North – East	0.477	0.345	0.493	0.277	0.401
South – East	0.404	0.247	0.528	0.244	0.332
South	0.426	0.257	0.471	0.279	0.340
South – West	0.395	0.287	0.546	0.273	0.352
West	0.280	0.225	0.383	0.164	0.257
North – West	0.313	0.228	0.319	0.162	0.262
Centre	0.255	0.208	0.324	0.170	0.234
Bucharest	0.149	0.117	0.419	0.075	0.159

Table 9.2. Multivariate analysis of the poverty degree

	Dependent variable: TFR index		
	Global	Housing conditions and durable goods	Financial difficulties
Intercept	0.10588	0.07126	0.42046
Occupational status of the household head: Employee (ref.)			
Employer	-0.07209	-0.05160	-0.21579
Self-employed in non-agricultural activities	0.05846	0.05076	0.06424
Farmer	0.13775	0.12973	0.11393
Unemployed	0.10093	0.07478	0.19935
Pensioner	0.01513	0.00583	0.05476
Other	0.11867	0.09954	0.15838
Educational level of the household head: High school (ref.)			
Primary or without school	0.18916	0.19730	0.08364
Lower secondary	0.10160	0.10237	0.07022
Vocational	0.03705	0.03174	0.05244
Post high school	-0.04101	-0.03472	-0.07043
University	-0.07812	-0.06120	-0.19751
Household type: Two adults with a child (ref.)			
Single man	0.04123	0.06544	-0.03819
Single woman	0.03261	0.04009	0.04639
Two adults without children	n.s.	0.01162	-0.04019
Other households without children	-0.00641	-0.00651	-0.05008
Two adults with two children	0.04697	0.00342	0.01886
Two adults with three and more children	0.13076	0.00477	0.09634
One adult with children	0.05306	0.00693	0.11115
Other households with children	0.05095	0.00306	0.01390
Area of residence: Urban (ref.)			
Rural	0.15365	0.19232	-0.07989
Region: Bucharest (ref.)			
North-East	0.07489	0.07794	0.03389
South-East	0.05137	0.04252	0.07489
South	0.02730	0.01827	0.02590
South-West	0.04892	0.03694	0.08943
West	n.s.	n.s.	-0.08789
North-West	-0.02320	-0.00904	-0.13325
Centre	-0.02648	-0.01831	-0.12176
R^2	0.5754	0.5959	0.1276

References

Betti G, Molnar M, Panduru F (2003) Consumption pattern and equivalence scales in Romania. Romanian Economic Review 12:207-221

CASPIS (2004) Trends of Poverty and Severe Poverty within 1995 - 2003. www.caspis.ro

Cheli B, Lemmi A (1995) A "Totally" Fuzzy and Relative Approach to the Multidimensional Analysis of Poverty. Economic Notes 24:115–134

Dinculescu, V, Chircă C, Câmpeanu M, Gheorghe D, Ivan-Ungureanu C, Molnar M, Panduru F, Pop AM (1999) Coordonate ale sărăciei în România. Dimensiuni şi factori (Poverty coordinates in Romania. Dimensions and factors). In: Sărăcia în România 1995-1998 (Poverty in Romania), UNDP, Proiectul de prevenire şi combatere a sărăciei, Bucureşti

INS, ICCV, IEN, MMPS, INCSMPS (1998) Methods and Instruments for Poverty Measurement, UNDP, Poverty Alleviation Project, Bucharest

Molnar M (1999) Sărăcia şi protecţia socială (Poverty and social protection). Bucureşti, Editura Fundaţiei "România de Mâine"

Panduru F, Molnar M, Gheorghe D (2000) Poverty in Romania, Seminar "International Comparisons of Poverty", SUSR, EUROSTAT, INSEE, Bratislava, 5-6 June

Panduru F, Molnar M, Pop L, Duma V (2001) Măsurarea sărăciei în România. Elemente pentru o nouă metodologie (Poverty measurement in Romania. A new methodology). Institutul Naţional de Statistică, CASPIS, Institutul de Economie Naţională, Universitatea Bucureşti, Bucureşti

Romanian Government – Anti-Poverty and Social Inclusion Commission (2002) National Anti-Poverty and Social Inclusion Plan

Romanian Presidency – National Commission for Preventing and Combating Poverty (1998) National Poverty Alleviation Strategy. Romania 1998

Teşliuc CM, Pop L, Teşliuc ED (2001) Sărăcia şi sistemul de protecţie socială (Poverty and social protection system). Iaşi, Polirom

UNDP (1999) Poverty in Romania, 1995-1998. Poverty Alleviation Project, Bucharest

World Bank (1997) Romania: Poverty and social Policy. Report No. 16462 – RO

World Bank (2003) Romania: Poverty Assessment. Report No. 26169 – RO

10 Multidimensional and Fuzzy Poverty in Switzerland

David Miceli

Département des Finances, Organisation et systèmes d'information

10.1 Introduction

During the last few decades, many attempts have been made to find a suitable way of measuring poverty. The first step is obviously to define poverty. There are of course many ways of defining poverty. For instance, in the absolute approach, some basic needs are taken into account for the poverty threshold (see Booth 1969; Rowntree 1901; Orshanski 1965; Watts 1967). An alternative approach is to define the poor relatively, by comparing the situation of each individual with the standard of living prevailing at a certain point in time in a given country (see Townsend 1979). Yet another approach, the subjective approach, lets the individuals evaluate their own situation (see Goedhart et al. 1977; Van Praag 1971).

These approaches have in common the fact that they all assess the poverty status of an individual or of a household by resorting to a unique indicator of resources, like income or expenditure. However, this procedure contains some drawbacks. In fact, each indicator reflects only a specific aspect of poverty. For instance, if we choose income as the relevant indicator for evaluating whether a person is poor or not, we assume that it gives us an idea of the opportunities that person has to meet some previously defined basic needs but, in no way do we know how the income is really spent. Moreover income alone does not tell us very much about an individual's living conditions. We can certainly admit that a relatively low level of income could be more than compensated for the fact that its recipient owns his house, for example.

Likewise, the use of expenditure as the indicator of resources is not entirely satisfactory either. Expenditure allows us to measure poverty from a standard of living point of view. But, again, we should not automatically consider people with lower consumption expenditures as poorer, because it can be the result of a choice consisting in selecting cheaper goods and services or, simply, not participating in certain activities.

Among the authors who advocate the use of alternative resource indicators that better reflect the living conditions of individuals, Travers and

Richardson recommend the use of direct measures of poverty along with full income. These measures are achieved by asking individuals how they evaluate their own situation in terms of food, clothes, shelter and transport, for example (see Travers and Richardson 1993).

Given the limitations related to measures of poverty based upon a single indicator, multivariate methods can be explored. With such techniques, various aspects of poverty can be included and summarised in a single number. This leads us to a much broader concept of poverty, reflecting other dimensions rather than just the monetary one. A major advantage of a multidimensional measure of poverty over the traditional ones is that it not only takes into account the material situation of individuals but it also captures their general living conditions. In addition, according to Whelan, a global index of poverty based on a set of deprivation indicators seems more appropriate than indices based only on income or expenditures to assess a situation of permanent poverty (see Whelan 1993). Such an index should ideally take into account basic needs, including food, clothing, housing and household equipment. Additionally it might also contain information on other variables, mostly related to social life and sometimes exerting some constraint on it. Working conditions, leisure, health, education, environment, family and social activities are some examples of these kinds of variables.

In a one-dimension framework, the current practice for measuring poverty is based on the assumption that the identification of the poor is possible with the use of a poverty line. Nevertheless it is difficult to achieve wide consensus on setting this limit. As pointed out by Cerioli and Zani or by Cheli et al., among others, the problem is in part due to the fact that a sharp division of the total population between poor and non-poor is unrealistic (see Cerioli and Zani 1990; Cheli et al. 1994). Mack and Lansley point out that "it is likely that there is a continuum of living standards from the poor to the rich, which will make any cut-off point somewhat arbitrary" (see Mack and Lansley 1985, p 41). One way of taking into account that characteristic is to take advantage of the tools provided by the theory of fuzzy sets. We agree with those authors who insist that if a notion is not exact by nature, we should not try to remove the degree of ambiguity it carries (see Basu 1987; Ok 1995, 1996).

10.2 Poverty in Switzerland

The methodology described by Cerioli and Zani has been applied to perform a multidimensional analysis of poverty in Switzerland (see Cerioli and Zani 1990). The data is obtained from the Consumption Survey conducted by the Swiss Federal Statistical Office for the year 1990. The Consumption Survey contains information on various categories of household expenditure, complemented with data on their income. On the other hand, precious data on a households' material conditions, such as the possession of durable goods or housing conditions are also available in the survey. The survey is based on a sample of households randomly selected from the Swiss phonebook. The sampling method presents some drawbacks. We can only rely on information concerning private households and therefore is restricted to those households with a telephone. In particular, the use of such a sample can lead to underestimating poverty; homeless people or elderly people living in institutions are among those who don't own a telephone.

The data was collected at two different levels. In the first part of the survey, households were asked to report, on an annual basis, general information on their expenditure. In the second part, much more detailed information was required from other households regarding their expenditures, but only for a given month. Only the first part is used in this analysis, as the data covers the same period for each household and is thus more comparable. After a coherence check of the data, a sample of 1963 observations remained.

To assess living conditions using a fuzzy measure of poverty, the indicators of deprivation need to be selected first. We have to keep in mind that the data we are using was designed for other purposes than that of measuring poverty. That is the reason why we have to limit ourselves to the variables available from the Consumption Survey. According to the Consumption Survey data, four categories of indicators were identified; housing conditions, the possession of durable goods, equivalent disposable income and equivalent expenditures[1]. The indicators associated with each category are presented in Table 10.1.

We now have to specify the form of the membership function for each indicator. From Table 10.1, we can easily see that we only have two categories of indicators. The ones labelled 1.1 and 2.1 to 2.8 are of the dichotomic type, while the remaining ones are continuous variables. For the dichotomic indicators, the membership function is obvious. An individual belongs to the subset of deprived people according to each dichotomic indicator, unless he is equipped with the good in question.

[1] Income and expenditures have been made comparable across households of different size and composition through the use of equivalence scales. The econometric scales estimated by Gerfin et al. with a Barten specification have been used (see Gerfin et al. 1994).

Table 10.1. Indicators of deprivation

1	Housing conditions
1.1	Hot water
1.2	Per capita square meters
2	Possession of durable goods
2.1	Cooker
2.2	Refrigerator
2.3	Deep-freeze
2.4	Dishwasher
2.5	Washing machine
2.6	Colour television
2.7	Video recorder
2.8	Car
3	Equivalent income
	Disposable income
4	Equivalent expenditures
4.1	Food
4.2	Clothing and footwear
4.3	Leisure, culture and hotels

On the other hand, we could choose several possible forms of membership functions for the continuous variables. However, we confine ourselves to a linear form which varies between two limits. The first one refers to the value of the chosen indicator of poverty defining some absolute poverty threshold, below which a given individual or household can without any hesitation be considered as poor. The second limit represents the value of the variable beyond which an individual can certainly be regarded as not poor[2]. For those values of the variable included between the two limits, the membership function must take its values in the interval [0,1].

Indicator 1.2 refers to the inhabitable area of the apartment. We suggest taking 25 square metres as the limit below which an apartment may be regarded as too small for a single person[3]. Accordingly, a person living in an apartment with a surface below this value faces extreme deprivation. On the other hand, we reckon that an individual residing in an apartment larger than the average size (50 square metres) is not deprived at all.

The second continuous indicator of deprivation found in Table 10.1 is given by equivalent disposable income. We propose setting the lower limit at half the value of a common poverty line – defined as half the median of

[2] The determination of the lower and upper bounds of such an interval is not always straightforward, because those limits generally depend upon the socio-economic context and on the specific characteristics of each indicator of privation.

[3] This limit is in accordance with the current standards in canton Geneva for the payment of housing subsidies.

the distribution – and the upper limit at twice the median equivalent disposable income[4]. By doing so, we judge that individuals with less than 9,308 Swiss francs per year belong entirely to the fuzzy subset of deprived people, while those with more than 74,464 Swiss francs per year may be considered as completely out of poverty.

Finally, for the last three continuous variables, related to equivalent expenditure, we simply decided to define the lower and upper limits as the minimum and maximum values encountered in the distribution, respectively. As a result, we find that the most deprived household, according to its food expenses, spent only 342 Swiss francs in 1990 and at the other extreme, the level of expenditure for food reached 17,085 Swiss francs for the only household totally outside of the fuzzy subset of deprived people. When we consider clothing and footwear expenditure, a minimum value of zero and a maximum value of 18,008 Swiss francs is observed. Finally, the minimum level of expenses for leisure, culture and hotels is 401 Swiss francs and the maximum 76,019 Swiss francs.

The results are presented in Table 10.2. Note that the average degree of belonging to poverty is very wide. We observe a minimum value of 0.0006, if we take the lack of hot water in the apartment as an indicator of deprivation. On the other hand, we find that the average degree of belonging to poverty reaches 0.9010 when we consider expenditure for leisure, culture and hotels as an indicator of deprivation. But we have to interpret these results carefully, especially when comparing values of average degree of belonging to poverty for different types of variables. Although it is clear that it represents a proportion for dichotomic variables, it is not necessarily true for continuous indicators.

Let us first analyse the results concerning the fuzzy proportion of poor households, according to indicators relating to the possession of durable goods. From Table 10.2, we notice that the cooker and the refrigerator are two widespread goods, since only 2.5% of Swiss households do not possess them. This result is not surprising, because those durable goods are used to store and transform food, which is a necessity. The average degree of deprivation is also low when we consider washing machines. From the Consumption Survey, we know that 7.1% of households do not own this

[4] The values taken by the two poverty lines might seem particularly low or high, for some of the continuous variables. The justification for these choices is that it is very likely that below the minimum threshold, individuals face bad enough living conditions to be considered as completely poor and beyond the maximum threshold they do not belong at all to the set of poor. In spite of that, the values chosen for the two thresholds for each continuous indicator of deprivation are completely arbitrary. Therefore a sensitivity analysis would be advisable, in order to investigate how the change of those limits would affect the results.

good. In this case, too, the presence of such a durable good amongst most of the households is not surprising, because it is related to the maintenance of clothing, which can also be regarded as a necessity.

Table 10.2. Fuzzy poverty in Switzerland

	Indicator of deprivation	Average degree of belonging to poverty	Weights
1	Housing conditions	0.0445	0.3356
1.1	Hot water	0.0006	0.2994
1.2	Per capita square metres	0.4076	0.0362
2	Possession of durable goods	0.1287	0.6140
2.1	Cooker	0.0258	0.1475
2.2	Refrigerator	0.0252	0.1484
2.3	Deep-freeze	0.4119	0.0358
2.4	Dishwasher	0.5924	0.0211
2.5	Washing machine	0.0714	0.1064
2.6	Colour television	0.1551	0.0752
2.7	Video recorder	0.6360	0.0182
2.8	Car	0.2181	0.0614
3	Equivalent income	0.5481	0.0242
4	Equivalent expenditures	0.7564	0.0262
4.1	Food	0.6778	0.0157
4.2	Clothing and footwear	0.8557	0.0063
4.3	Leisure, culture and hotels	0.9010	0.0042
	Fuzzy index of poverty	0.1270	

The results found for the other items belonging to the second group of indicators give us more information on the lifestyle of households and on their more or less well-off living conditions. Thus it appears that almost 16% of households do not possess a colour television and 22% do not have a car. Finally, we can suppose that the possession of the remaining durable goods is more a question of taste. Nevertheless, owning one of those additional items probably gives the households better living conditions. More than half of Swiss households do not have a dishwasher or a video recorder at their disposal, while only 59% have a deep-freeze.

Let us now turn to indicators of deprivation related to equivalent expenditures of households. The interpretation of the fuzzy proportion of poor households is not as straightforward as in the case of possession of durable goods. Here, we should consider the average degree of belonging to poverty as the average relative position of households in relation to two extreme situations; that of the most deprived and that of the best-off. Keeping this interpretation in mind, we note that the equivalent expenditures are on average closer to the bottom end of the distribution[5]. As we would expect, the average deprivation is lowest for food expenditure and the highest when we assess living conditions by referring to expenditure for leisure, culture and hotels. In fact, we often observe less dispersion in expenditures for goods that can be considered as necessities.

In the last column of Table 10.2 the weights associated with the different indicators of deprivation can be found. As can be seen, those related to the possession of durable goods receive the highest weight (61.4%), followed directly by those related to the housing conditions (33.6%). On the other hand, the monetary indicators of deprivation, concerning equivalent income, as well as expenditure, account for only 5% in the aggregation process. This is mainly due firstly to the definition of the membership functions, which tend to give higher deprivation scores for this kind of variable, and secondly to the way the weights are made dependent on the average level of deprivation. The choice of both the membership functions and the weighting system has a strong impact on the overall fuzzy index of poverty. This should guide us towards interpreting the overall fuzzy index of poverty from a relative point of view, when comparing living conditions through time or over different population subgroups. Thus the value of 0.1270 in the last row of Table 10.2 does not have any particular meaning in the absolute. In other words, it would be very difficult to say whether this value reflects good or bad living conditions. The above assertion leads us naturally to perform some decomposition of the global fuzzy index by selecting subgroups of population.

10.3 Decompositions of poverty

In this section, we present two decompositions of the fuzzy index of poverty. For the first decomposition, the population is divided into five subgroups according to the employment status of the head of household[6]. For the second, seven groups are distinguished according to the household's composition. For each subgroup considered, their overall level of poverty is calculated by using the same set of weights during the process of aggregation of the indicators of deprivation. By doing so, the fuzzy index of poverty for the whole population is obtained as a weighted average of the

[5] This result is not surprising because of the typical asymmetrical and right-skewed form of the distribution, combined with the fact that extreme observations are replaced with values closer to the centre of the distribution.

[6] The head of the household is defined in the Consumption Survey as the person who financially contributes the most to the total income of the household.

observed poverty in each subgroup, the weights being given by the shares of the population. We could, of course, calculate the weights separately within each subgroup, in order to take their peculiarities and the relative importance of each indicator of deprivation into account. As a matter of fact, some groups in the population might have different needs and tastes and accordingly don't give the same importance to the selected indicators. However, we prefer using the same system of weights for each subgroup because it makes it easier to compare the level of deprivation between the groups.

10.3.1 Poverty by employment status

For the first decomposition of the fuzzy index of poverty, we have five population subgroups, each one related to a different employment status. In the first category, we find households of which the head is self-employed. This category essentially includes traders, entrepreneurs and persons practising a liberal profession. The self-employed represent 6.8% of the whole population. Farmers are not part of this category and form the second subgroup. Their share in the total population is only 4.7%. In the third subgroup, we find households with an employee as their head. This category is quite wide and heterogeneous, as it contains directors or state employees, junior and senior executives, employees, workers, trainees, as well as apprentices. This group accounts for a large proportion, 64.1%, of the population. The second highest share is obtained for the fourth subgroup, made up of pensioners, which reaches 21.9%. Finally, the last category comprises households with a head that has any other activity and includes unemployed people or students, for example. This last subgroup is the smallest one, and represents only 2.6% of the whole population.

When analysing the results presented in Table 10.3, we see that employees enjoy the best living conditions, according to the fuzzy index of poverty. They are directly followed by farmers and the self-employed. For those three subgroups, the level of poverty is less than the national average of 0.1270 and the fuzzy index of poverty takes respectively the value of 0.1186, 0.1201 and 0.1255. Then households where the head is retired seem to be more deprived, as their average deprivation reaches 0.1474. Finally, the last subgroup presents the highest degree of membership to the fuzzy subset of the poor, with an average value of 0.1811.

It is extremely interesting to compare those results with the ones obtained using other approaches for measuring poverty. More specifically, in previous work we used the headcount ratio with a poverty line set at half the median of the distribution employing both disposable income and consumption expenditures as an indicator of resources[7]. As expected, the ranking of the five subgroups differs slightly, depending on the choice of the indicator (see Miceli 1997, pp. 143-148).

Table 10.3. Decomposition of fuzzy poverty by employment status

Group[8]	(1)	(2)	(3)	(4)	(5)	Total
%of pop.[9]	6.8%	4.7%	64.1%	21.9%	2.6%	100%
Ind.[10]	Average degree of belonging to poverty					
1	0.0477	0.0410	0.0512	0.0233	0.0542	0.0445
1.1	0.0000	0.0000	0.0009	0.0000	0.0000	0.0006
1.2	0.4420	0.3805	0.4675	0.2158	0.5025	0.4076
2	0.1225	0.1091	0.1150	0.1666	0.2013	0.1287
2.1	0.0447	0.0127	0.0191	0.0386	0.0578	0.0258
2.2	0.0655	0.0085	0.0196	0.0323	0.0289	0.0252
2.3	0.3994	0.1098	0.4172	0.4650	0.4102	0.4119
2.4	0.3338	0.5556	0.5758	0.7202	0.6748	0.5924
2.5	0.0684	0.0078	0.0687	0.0766	0.2170	0.0714
2.6	0.1783	0.3473	0.1484	0.1113	0.2838	0.1551
2.7	0.5297	0.8412	0.5658	0.8157	0.7706	0.6360
2.8	0.1177	0.0962	0.1467	0.4657	0.3807	0.2181
3	0.6426	0.8958	0.4600	0.6779	0.7607	0.5481
4	0.7164	0.6728	0.7507	0.7984	0.7976	0.7564
4.1	0.6285	0.5013	0.6789	0.7223	0.7232	0.6778
4.2	0.8201	0.9050	0.8381	0.9035	0.8959	0.8557
4.3	0.8896	0.9658	0.8881	0.9254	0.9284	0.9010
FIP[11]	0.1255	0.1201	0.1186	0.1474	0.1811	0.1270

The main difference we observe, when we use a unique indicator of deprivation for measuring poverty, is the change in the relative position of the self-employed. While they seem to be rather deprived, according to their disposable income level, they figure amongst the best-off if we examine their consumption expenditures. This is probably due to the way self-employed households report their income, clearly tending to underestimate it. But when we include indicators other than just those related to the monetary situation of households, we discover that the living conditions of farmers are also not as bad as a one-dimension analysis would suggest.

[7] This definition of the poverty line corresponds to a level of disposable income amounting to 18,616 Swiss francs. If we use consumption expenditures as the indicator of resources, the poverty line takes the value of 17,503 Swiss francs, according to the above mentioned definition.

[8] (1) Self-employed; (2) Employed in the agricultural sector; (3) Employee; (4) Retired; (5) Other.

[9] Share of the population.

[10] Indicator of deprivation.

[11] Fuzzy index of poverty.

Furthermore, retired households, but above all unemployed people or students, display a worsening of their situation when we include more material conditions. The image provided by a multidimensional analysis is definitely closer to the common idea of poverty and its intensity in the different groups of households, according to their employment status.

Let us now analyse in more detail the results presented in Table 10.3. If we focus on each group of indicators separately, we do not systematically find the same ranking of households by employment status as the one obtained for the overall fuzzy index. The only thing common to all the groups of indicators, when we assess living conditions, is that unemployed people or students always appear amongst the households with the highest deprivation levels and the self-employed are always relatively well-off.

We notice that retired households enjoy the best housing conditions from the net floor area point of view. This can in part be explained by the fact that they often live in older apartments, which tend to be more spacious. Also in most cases they stay in the same apartment most of the time, even when children leave home.

When we evaluate living conditions by considering the possession of some durable goods, we note that on average farmers experience the lowest deprivation level. They are often better equipped than other households, especially with durable goods directly related to basic needs like food and clothing. In fact, the rates of deprivation observed for the possession of a cooker, a refrigerator, a deep-freeze and a washing machine are markedly lower than the national average. This is probably due to the fact that most farmers are isolated from urban centres, which forces them to possess those items. Moreover, less than 10% of the households employed in the agricultural sector do not own a car. Again this represents the lowest deprivation rate. A plausible explanation is that distances are longer and public transport is less developed in rural than in urban areas. In contrast, durable goods related to leisure, such as a colour television or a video recorder, are definitely less frequent amongst farmers than in other types of households. The main characteristic of self-employed living conditions lies in a moderate deprivation in almost all durable goods, with the exception of the cooker and the refrigerator. They even figure at the first rank for dish-washers or video recorders. As for the retired, they tend to possess the various durable goods to a lesser extent than the rest of the population, which is quite understandable when we consider the ownership of a car, for example. On the other hand, in this group we find the highest percentage of households possessing a colour television.

When we use disposable income to assess living conditions, the ranking of households remains almost unchanged compared to the one obtained with the group of indicators for the possession of durable goods, with the exception of the relative position of farmers. They are indeed less deprived with regard to their possession of various durable goods, whereas their apparent deprivation reaches the highest level in terms of disposable income.

Finally, when we estimate the living conditions of households by resorting to their expenditures on selected goods and services, we obtain more or less the same image as the one depicted by the absence of durable goods as an indicator of deprivation. Thus, retired and unemployed people, as well as students, face the worst living conditions. However, we notice that the differences in the level of deprivation for the different groups seem very small compared to the ones observed when using other groups of indicators. It could then be worth checking whether the differences are statistically significant.

10.3.2 Poverty by household composition

For this second decomposition of the fuzzy index of poverty, the households by composition are distinguished. By doing so, we can investigate if the size and composition of households can influence their living conditions. Seven subgroups of the population have been defined. In the first category consisting of single adults, we find 24.9% of households. Single parents form the second subgroup. Their share in the population is only 2%. The next subgroup is the largest one, containing 36.5% of households with two adults without children. The percentage of households made up of two adults and one child reaches 10.3%. Then 13.1% of households are found among couples with two children. The next group includes couples with more than two children. This group accounts for 7.1% of the population. Finally, the proportion of households with more than two adults is 6.0%.

The fuzzy proportions of poor households, according to the various indicators or groups of indicators, are presented in Table 10.4. From the last row, we observe that living conditions tend to improve with the number of adults in the household. Households with more than two adults can be considered as the ones which face the best living conditions. Their average degree of belonging to poverty is only 0.1025. Moreover, all the households consisting of two adults, with or without children, show overall scores of poverty inferior to the Swiss national average of 0.1270 or at least close to that value. On the other hand, households with just one adult face worse living conditions, especially with the presence of children. As can be seen from the first row of Table 10.4, we find a similar result when we consider poverty from the housing conditions viewpoint. In fact, they tend to deteriorate with the size of household.

Table 10.4. Decomposition of fuzzy poverty by household composition

Group[12]	(1)	(2)	(3)	(4)	(5)	(6)	(7)	Total
% of the pop.[13]	24.9%	2.0%	36.5%	10.3%	13.1%	7.1%	6.0%	100%
Ind.[14]			Average degree of belonging to poverty					
1	0.0133	0.0681	0.0324	0.0617	0.0844	0.0884	0.0701	0.0445
1.1	0.0024	0.0000	0.0000	0.0000	0.0000	0.0000	0.0000	0.0006
1.2	0.1040	0.6313	0.3006	0.5724	0.7832	0.8201	0.6502	0.4076
2	0.1975	0.1870	0.1190	0.0977	0.0770	0.1013	0.0715	0.1287
2.1	0.0341	0.0627	0.0625	0.0162	0.0104	0.0245	0.0267	0.0258
2.2	0.0292	0.0823	0.0307	0.0113	0.0148	0.0191	0.0099	0.0252
2.3	0.6465	0.4215	0.4059	0.3391	0.2290	0.1633	0.2921	0.4119
2.4	0.7869	0.6158	0.5958	0.4595	0.4711	0.4272	0.4465	0.5924
2.5	0.1154	0.1553	0.0626	0.0617	0.0328	0.0434	0.0483	0.0714
2.6	0.2176	0.1507	0.1243	0.1653	0.0981	0.2526	0.0747	0.1551
2.7	0.7592	0.7322	0.6159	0.5405	0.5460	0.6721	0.5324	0.6360
2.8	0.4834	0.3925	0.1674	0.0851	0.0748	0.0816	0.0704	0.2181
3	0.5975	0.6618	0.5117	0.4957	0.5544	0.6818	0.4441	0.5481
4	0.8290	0.7665	0.7480	0.7244	0.7123	0.7220	0.6940	0.7564
4.1	0.7871	0.6898	0.6690	0.6301	0.6076	0.6116	0.5878	0.6778
4.2	0.8805	0.8591	0.8503	0.8421	0.8426	0.8585	0.8339	0.8557
4.3	0.9086	0.9142	0.8902	0.9002	0.9086	0.9298	0.8813	0.9010
FIP[15]	0.1619	0.1738	0.1159	0.1117	0.1077	0.1273	0.1025	0.1270

Let us now focus on living conditions using the possession of durable goods as the indicator of deprivation. It is striking to note that households with less than one adult are the only ones to show a higher degree of poverty than the national average. The possession of durable goods seems to increase with the size of households. If we now turn to deprivation levels, as depicted by disposable income, we find that households with more than two adults enjoy the best living conditions.

On the other hand, single parents and numerous families seem to figure amongst the most deprived from the disposable income point of view.

Finally, the ranking of households obtained when we use expenditure as the indicator of deprivation is almost the same as the one provided by the use of disposable income or durable goods.

[12] (1) 1 adult; (2) 1 adult with children; (3) 2 adults; (4) 2 adults with 1 child; (5) 2 adults with 2 children; (6) 2 adults with more than 2 children; (7) more than 2 adults.

[13] Shares of the population.

[14] Indicator of deprivation.

[15] Fuzzy index of poverty.

10.4 Concluding remarks

This Chapter presents an example of the application of multidimensional measurement of poverty, using fuzzy sets. The empirical results obtained for Switzerland in 1990 show that the use of several indicators not only helps in giving a more complete picture of living conditions, but also gives us an image of poverty closer to what we could perceive by just observing reality.

When comparing poverty between groups of the population defined according to the employment status of the head of household, we find the highest degrees of deprivation among the retired, the unemployed and students. The fuzzy index of poverty appears to be a superior measure to the headcount ratio. In fact, although farmers in general have a low level of disposable income and of consumption expenditures, that does not mean that they face worse living conditions than the rest of the population. Actually the fuzzy index of poverty shows that when we take account of the housing conditions and the possession of durable goods, farmers do rather well.

The results obtained from the decomposition of overall poverty by household composition indicate that households with less than two adults present highest levels of deprivation. Again results obtained from the fuzzy approach seem more reliable than the ones provided by the headcount ratio where single parents figure amongst the less deprived. That does not necessarily correspond to the observed reality.

The analysis of poverty presented in this Chapter is essentially based on an external observer's point of view of what deprivation represents. The fact of not possessing a certain good or of having less than other groups of the population is supposed to increase the sense of deprivation. However the subgroups identified as belonging to the most deprived might not think they do. Our analysis would certainly be improved if we could include indicators reflecting how the households evaluate their own situation. As an example, households with a retired head may not find it necessary to possess a video recorder.

Obviously, the results depend strongly upon the choice of indicators for this kind of analysis so it would be worth defining an appropriate set of indicators to include in the data, covering all the relevant areas for an analysis of living conditions.

From the theoretical and conceptual point of view, we could investigate different methods of aggregation and weighting systems. Given that both issues are subject to controversy, the evaluation of living conditions using the fuzzy index of poverty should go hand in hand with a sensitivity analysis on the form of the membership function for continuous variables as well as a sensitivity analysis on the system of weighting.

Furthermore, the results obtained are subject to sample variability. It would then be interesting to take account of these possible variations in the technique, to see whether the levels of deprivation are statistically different from one another. This would be very useful, especially when dealing with small samples.

Therefore, we strongly conclude that this kind of multidimensional approach for measuring poverty is much more realistic than the traditional ones, based on a single indicator of resources. Although the interpretation of the final index is not very easy, because it combines indicators of a different nature, we can still have an insight into the major areas of living conditions by separately considering groups of indicators, like housing conditions, the possession of durable goods, working conditions, and so forth. The overall fuzzy index of poverty is also shown to be particularly illuminating when comparing several groups of the population.

References

Basu K (1987) Axioms for a Fuzzy Measure of Inequality. Mathematical Social Sciences 14:275-288

Booth C (1969) Life and Labour of the People of London. A. M. Kelley, New York

Cerioli A, Zani S (1990) A Fuzzy Approach to the Measurement of Poverty. In: Dagum C, Zenga M (eds) Income and Wealth Distribution, Inequality and Poverty. Studies in Contemporary Economics. Springer Verlag, Berlin, pp 272-284

Cheli B, Ghellini G, Lemmi A, Pannuzi N (1994) Measuring Poverty in the Countries in Transition via TFR Method: The Case of Poland in 1990-1991. Statistics in Transition 1:585-636

Gerfin M, Leu RE, Schwendener P (1994) Ausgaben-Äquivalenzskalen für die Schweiz: Theoretische Ansätze und Skalen aufgrund der Verbrauchserhebung 1990. Technical report, Office federal de la statistique, Bern

Goedhart T, Halbertstadt V, Kapteyn A, Van Praag BMS (1977) The Poverty Line: Concept and Measurement. The Journal of Human Resources 12:503-520

Mack J, Lansley S (1985) Poor Britain. Allen and Unwin, London

Miceli D (1997) Mesure de la pauvreté: Théorie et application à la Suisse. Ph.D. thesis, Université de Genève

Ok EA (1995) Fuzzy Measurement of Income Inequality: A Class of Fuzzy Inequality Measures. Social Choice and Welfare 12:111-136

Ok EA (1996) Fuzzy Measurement of Income Inequality: Some Possibility Results on the Fuzzification of the Lorenz Ordering. Economic Theory 7:513-530

Orshanski M (1965) Counting the Poor: Another Look at the Poverty Profile. Social Security Bulletin 28:3-29

Rowntree BS (1901) Poverty: A Study of Town Life. MacMillan, London

Townsend P (1979) Poverty in the United Kingdom. Penguin Books, Middlesex

Travers P, Richardson S (1993) Living Decently: Material Well-Being in Australia. Oxford University Press, Melbourne

Van Praag BMS (1971) The Welfare Function of Income in Belgium: An Empirical Investigation. European Economic Review 2:337-369

Watts HW (1967) The Iso-Prop Index: An Approach to the Determination of Differential Poverty Income Thresholds. The Journal of Human Resources 2:3-18

Whelan BJ (1993) Non Monetary Indicators of Poverty: A Review of Approaches. Paper presented at the Conference on Household Panel Surveys, Luxembourg, June 1-2, 1993

11 A Comparison of Poverty According to Primary Goods, Capabilities and Outcomes. Evidence from French School Leavers' Surveys

Josiane Vero

Département 'Production et Usages de la Formation Continue', University of Marseille

11.1 Introduction

There has been a lively debate on the nature and the definition of poverty. Most experts have long understood that poverty is inclined to vary through time and space. Poverty has a somewhat different connotation today in France from what it has in the developing countries or from it had in the past. That having been said, poverty is a difficult notion and it may be defined in various ways which correspond to different philosophical approaches. The general idea is that poverty is a consequence of an inequality, between individuals, in the control of certain things, i.e., the result of an unequal distribution between those who have something and those who are more or less deprived by it. Poverty is then a situation in which certain individuals are deprived of this something. Thus, according to Sen, the central question to define and measure inequality, as well as poverty can be resumed as follows: "equality of what?" (Sen 1980, 1987b). Thus, in order to define and measure poverty one has to formulate a value judgement on what must be the objects of value. Discussions on normative economics have offered us a wide menu in answer to this question "equality of what?": for example, income, wealth, rights, freedom, etc. In this chapter, I shall concentrate on three particular types of responses that specify the objects of value for equality and poverty, which may be called the informational base.

- Approaches using means of freedom (Rawls 1971)
- Approaches shifting attention from means to what means do to human beings (Sen 1985)
- Approaches selecting social outcomes (Fleurbaey 1995).

I shall argue that the selection of normative principles on the definition of a poverty concept has a strong impact on the population of the individuals to be considered poor. The first part is dedicated to the presentation of three approaches of liberal egalitarianism. The central place of John Rawls

Theory of Justice (1971) is impossible to circumvent both because of its originality and because of the influence it had on subsequent work. The second approach, proposed by Sen focuses on the informational basis of the notion of capability (Sen 1985). The third approach, taken by Fleurbaey (1995) stresses the informational basis of the concept of social outcomes. The second part of the paper will look at the implications of the previous distinctions for poverty measurement. All three points of view, stressing respectively primary goods, social outcomes and capabilities, suggest resorting to a multidimensional approach. Several suggestions have been made in the past to take a multidimensional view of inequality and poverty and we have decided to adopt what has been called the fuzzy approach to poverty measurement (Cerioli and Zani 1990). The basic idea is to reject the notion of a simple binary choice (being poor or not) and to admit, on the contrary, that in many cases there are intermediate situations. The third part of the paper will give an illustration of the choice of informational space and of its consequences. The three competitive approaches, that have been mentioned previously, will be tested on the basis of data (1999) collected by the French Centre of Research in Education, Training and Employment (CEREQ). We conclude that the results derived from the social outcomes and capability approaches are often similar whereas a focus on primary goods identifies a totally different population

11.2 Three concepts of poverty

In this section, we have two aims: first, we shall be concerned with clarifying basic features of the primary good approach, (Rawls 1971), the capability approach (Sen 1985) and the social outcomes approach (Fleurbaey 1995); and second with explaining the connections between the three concepts.

11.2.1 Clarifying basic features

In an exercise of evaluation, a central question will be distinguished: what are the objects of value? The identification of the objects of value specifies what may be called the informational base for the measurement of poverty. Consequently answering the question about the objects of value provides information about what the relevant informational base does include and what it excludes in order to evaluate poverty. It's also necessary to clarify basic features.

On social primary goods

First, we shall present the notion of social primary goods, a notion which is part of the conception of justice presented in Rawls book: *A theory of Justice* (Rawls 1971). Rawls himself says that social primary goods are "things that every rational man is presumed to want". Consequently, he uses social primary goods as the index of advantage. These primary goods may be characterised under five headings as follows: First, the basic liberties are given by a list, for example: freedom of thought and liberty of conscience; freedom of association; and the freedom defined by liberty and integrity of the person, as well as by the rule of law; and finally political liberties; Second, freedom of movement and choice of occupation against a background of diverse opportunities; Third, powers and prerogatives of offices and positions of responsibility, particularly those in the main political and economic institutions; Fourth, income and wealth; and finally, the social bases of self-respect. It's interesting at this point to understand that primary goods are to be used in making comparisons for questions of social justice. An index of primary goods defines a public basis of interpersonal comparisons for questions of social justice. We are required to examine citizen's level on primary goods and furthermore an individual index of social primary goods is to be used in order to evaluate poverty. What is crucial for the problem under poverty is the concentration on bundles of primary goods. Rawls justifies this in terms of a person's responsibility for his own ends.

On capabilities

Sen (1992) criticizes Rawls' views and offers his own answer, which is that people should be equal in their *capabilities*. He considers that the primary goods approach takes little note of the diversity of human beings. People are not similar. They have different needs varying for example with health, longevity, climatic conditions, temperament, and even body size (affecting food and clothing needs). So judging advantage purely in terms of primary goods implies that individuals have the same needs and that they have full control over the conversion of primary goods in functionings. So Sen's view is that the quality of a person's life should be assessed in terms of the person's capabilities. A capability is the ability or potential to do or be something, more technically to achieve a certain functioning. Functionings represent parts of the state of a person, in particular the various things that he or she manages to do or be in leading a life. The capability of a person reflects the alternative combinations of functionings the person can achieve and from which he or she can choose one collection.

Sen's view is that people ought to be made equal in their capabilities or at least in their basic capabilities.

The corresponding approach to poverty takes the sets of individual capabilities as constituting an indispensable and central part of the relevant informational base of such an evaluation. It differs from other approaches using means of freedom focusing on the primary goods such as in a Rawlsian theory of justice. For example, the capability approach differs from the views of Rawls in making room for a variety of human acts and states as important in themselves. On the other hand, the approach does not attach direct importance to the means of freedom (primary goods, resources), like Rawls' approach does. For poverty evaluation it may be useful to identify a subset of crucially important capabilities dealing with what have come to be known as "basic needs". The term "basic capabilities" used in Sen (1980) was intended to separate out the liability to satisfy certain crucially important functionings up to certain minimally adequate levels. The identification of minimally acceptable levels of certain basic capabilities (below which people are considered as being scandalously deprived) provides an approach to poverty. Basic capabilities concerns for example: "The ability to move, to meet one's nutritional requirements, the wherewithal to be closed and sheltered, the power to participate in the social life of the community" (Sen 1987b). But one can consider others. According to Sen's point of view, capabilities vary between time and between communities at the same time. That's why Sen rejects the idea of giving a canonical list of basic capabilities. Another reason for such a position is provided by the necessity of a social debate. I will conclude this presentation with a pragmatic remark. That having been said, there are many formal problems involved in the evaluation of poverty based on capabilities, because of all of the combinations of functionings which are possible for an individual, i.e. capabilities can not be observed. It is in fact only possible to characterize functionings in a "refined" way to take into account the counterfactual opportunities. Corresponding to the functioning x, a "refined functioning" takes the form of "having functioning x through choosing it from the set S".

On social outcomes

On a paper headed "Equal Opportunity or Equal Social outcome", Fleurbaey (1995) has provided a critical assessment of Sen's writings on capability and of Rawls writings on primary goods, at the same time presenting his own answer to the question "equality of what?" Fleurbaey's main thesis is that Rawls neglects ability differentials, which are unjust inequalities amongst individuals. But with Rawls and against Sen, it argues that social institutions should not take care of the fate of individuals in a comprehensive way. Only the distribution of social outcomes, which might also be named "primary functionings", matters from the standpoint of social justice and poverty measurement. An application of a social outcomes approach which would seem plausible to him for western societies would select only six individual outcomes: respect for the private sphere, health, education and information, wealth, collective decision making power, and social integration. But at the same time Fleurbaey explains that "*this is just an example, and of course many details have to be worked out concerning the measurement of these six variables.*" That having been said, a series of specific objections will be given. But the main problem is that the philosophical basis of social outcomes neglects freedom to choose the relevant outcomes that people have reasons to promote.

To avoid confusion, it should be noted that the concept of social outcomes is used in a different sense from the concept of functionings. The main difference concerns subjective indicators, which are excluded from social outcomes whereas informational base of functionings has both subjective and objective features.

11.2.1 Describing connections between the three concepts

To continue this section focus will be on the links between the three concepts and on the approaches to poverty in their responsibility-based version. A graph is used to explain the links between primary goods, capabilities and social outcomes. There are two important links in the chain from primary goods to primary social outcomes. These links are summarized in Figure 11.1 inspired by Muellbaeuer's analysis (Muellbauer 1987).

First, primary goods are transformed into capabilities of a person to function, for example to be well nourished. These primary goods may translate in different ways. Apart from personal details, some other characteristics influence the capabilities of a person to function such as physical, social and political environment. Secondly capabilities of an individual as well as his psychic state for determining the levels of achievement in the different types of social outcomes.

Of course, each concept of poverty tells us about the way the society deals with its responsibility (Fleurbaey 1998) and with an individual's responsibility. When poverty is based on primary goods, society's responsibility is assigned over some means of freedom. To make up for it, the individual is left to his own means to define personal goals and ambitions and to transform primary goods into human beings. On the other hand, in calculating poverty on capabilities, Sen's principle sets the cut between choices. Society's responsibility concerns a set of capabilities, i.e. a combination of functionings and in return individuals have ultimate control over them. Consequently, an individual is responsible for choosing one collection of functionings and for achieving them. Finally, adopting social outcomes as objects of value in the measurement of poverty suggests that society decides on a bundle of functionings it considers important and his responsibility consists in guaranteeing them for all the individuals. In exchange, the responsibility for achieving other functionings of minor importance is attributed to individuals.

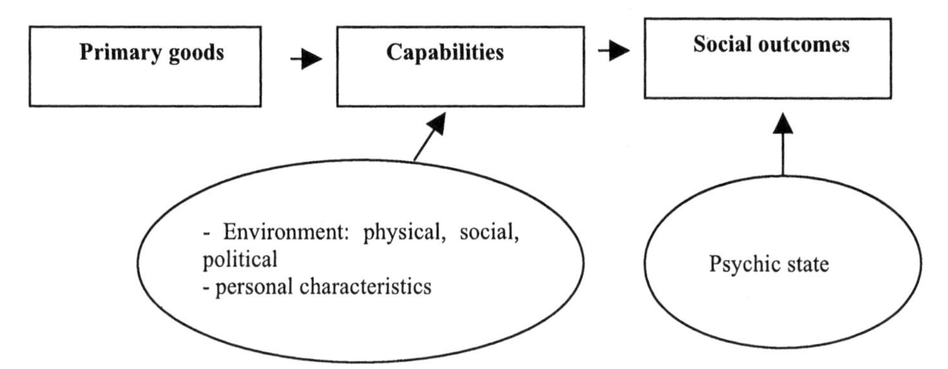

Fig. 11.1. Links between primary goods, capabilities and social outcomes

11.3 A multidimensional measure of poverty: the fuzzy logic

In this section we shall propose methodological tools to measure and compare the three conceptions of poverty. The problem is twofold. The first problem is to measure poverty in a multidimensional framework. In addition, the second problem is to go further than a binary vision of poverty restoring the individual situations in terms of poor and non-poor persons.

In order to take into account these two problems we shall propose a multidimensional new measure of poverty based on fuzzy logic (Vero and Werquin 1997; Vero 2002). Zadeh (1965), followed by Dubois and Prade (1980) introduce the fuzzy set theory, which is the starting point of view of our own study. Fuzzy sets are mathematical tools, which allow for the identification of objects, which do not have membership accurate criteria. A fuzzy system allows a gradual and continuous transition, say, from 0 to 1, rather than a crisp and abrupt change between binary values of 0 and 1.

To be concrete, consider first the ordinary sets principle, which is prevalent in measuring poverty. Let X be a set and x an element of X. Let A be a subset of X. The element x can take two different positions as regard to A:

$$\begin{cases} x \in A \Leftrightarrow \mu^*_A(x) = 1 \\ x \notin A \Leftrightarrow \mu^*_A(x) = 0 \end{cases} \tag{11.1}$$

Where μ^*_A is the membership function. One can view the traditional measure of poverty as deriving from an interpretation of the ordinary sets principle. That interpretation says that all individual i (i=1, ..., n) in the population N are classified in the poor subset P, of N, according to the following criterion:

$$\begin{cases} y_i < Z \Leftrightarrow i \in P; \quad \mu^*_P(i) = 1 \\ y_i \geq Z \Rightarrow i \notin P; \quad \mu^*_P(i) = 0 \end{cases} \tag{11.2}$$

Consider now the fuzzy sets principle. Let X be a set and x an element of X. A fuzzy subset A of X is defined as follows for all x belonging to X:

$$\{x, \mu_A(x)\} \tag{11.3}$$

where μ_A is a membership function which takes its values in the closed interval [0;1]. Each value $\mu_A(x)$ is the degree of membership of x to A. Consequently, the element x can take three different positions as regard to A:

$$\begin{cases} \mu_A(x) = 0 \\ 0 < \mu_A(x) < 1 \\ \mu_A(x) = 1 \end{cases} \tag{11.4}$$

Thus, if $\mu_A(x)=0$ then x certainly does not belong to A. If $\mu_A(x)=1$, then x completely belongs to A and if x is such as $0<\mu_A(x)<1$ then x partially belongs to A and its degree of membership is given by the value of $\mu_A(x)$.

To continue with the poverty application, note that in a fuzzy approach the membership function to the poor set of individual i is defined as follows: (i) the membership function is zero if the individual is certainly non poor; (ii) is between zero and one if the individual reveals only a partial membership to the fuzzy set of poor; (iii) is one if the individual completely belongs to the set.

Although we have just explained how the fuzzy logic consists in measuring poverty, the main issue of this approach is to specify a membership function. In practice, there are numerous fuzzy membership functions in the domain of poverty, which allow us to represent qualitative or continuous variables. Nevertheless, most of them are based on the work of Cerioli and Zani (1990) from which fuzzy poverty measures have been successfully developed (Cheli 1995; Cheli et al. 1994; Cheli and Lemmi 1995).

Clearly our poverty fuzzy measure refers to the work of Cerioli and Zani (1990), which initiated a statistical method for multidimensional analysis in which poverty is treated as a fuzzy concept, liable to assume a variety of shades and degrees, but we have developed a fuzzy method, which deviates from the initial proposition. In fact, we adopted their manner to define the membership function of the income, or qualitative indicator or continuous variable. Nevertheless we don't agree with the manner of summing up and weighting all the indicators selected. In fact, the basic problem of this approach is how it sums up the many facets of individual poverty and emphasizes the different degrees to which each subject may be regarded as poor. Indeed, the weight assigned to each indicator of poverty variable X is determined independently of the possible correlation with another indicator of poverty Y. Consequently, a difficulty arises from their proposition because it avoids excessive importance being assigned to correlated indicators and redundant variables. To solve the problem, we need another weighting, based not only on the occurrence of an indicator but also on the occurrence of a vector of variables. We attempted to propose a precise way that can minimize the relative weight of redundant indicators and rebalance the weighting of correlated variables (Vero and Werquin 1997; Vero 2002).

Let us first of all present previous studies on which our data processing is based. We turn next to our proposed membership function based on an alternative weighting.

11.3.1 Data processing: income, qualitative and continuous indicators

In this section, we shall present a data processing of income, and non-monetary variables as Cerioli and Zani (1990) first used in the context of poverty. We first consider the case where total income y_i of the *ith* individual is known. The membership function to the poor is then defined by fixing a value z' up to which an individual i is definitively poor and a value z" above which an individual i is definitively non-poor. Thus, we have:

$$\mu_j\left(x_{ij}\right)=\begin{cases} 1 & \text{if} \quad 0<y_i<z' \\ \dfrac{z''-y_i}{z''-z'} & \text{if} \quad z' \geq y_i > z'' \\ 0 & \text{if} \quad y_i \geq z'' \end{cases} \tag{11.5}$$

Following Cerioli and Zani (1990), we secondly consider indicators complementary to income in a multidimensional framework. Suppose that k variables $X_1, X_2, ..., X_k$ are observed on the n individuals of the population and let x_{ij} denote the level of a variable X_j (j=1,2, ..., k) observed for the *ith* individual (i=1,2, ..., n). As introduced in Chapter 3, Cerioli and Zani (1990) considered a transition zone $x_j^L < x_{ij} \leq x_{ij}^H$ for attribute j over which the membership function declines from 1 to 0 linearly:

$$\mu_j(x_{ij}) = \begin{cases} 1 & \text{if} \quad x_{ij} \leq x_j^{(L)} \\ \dfrac{x_j^{(H)} - x_{ij}}{x_j^{(H)} - x_j^{(L)}} & \text{if} \quad x_{ij} \in \left(x_j^{(L)}, x_j^{(H)}\right] \\ 0 & \text{if} \quad x_{ij} > x_j^{(H)} \end{cases} \tag{11.6}$$

Then, they identify the poor people as those who are excluded from the common standard of living. Individuals deprived of widespread commodities and who have a life style below the standard of the population. The aim is to build an index which goes beyond the income and which allows all the dimensions of the situations of poor people to be captured. We shall not go into detail about the construction of their membership function but we just want to stress the distinction between their approach and our proposed one.

Cerioli and Zani (1990), for each individual i, use the value of an indicator X_1, which represents for instance whether the commodity 1 is owned or not, as compared to the distribution of X_1 among the population. The more the commodity 1 is widespread among the population, the higher the deprivation for individual i and the higher the weight for indicator X_1 in the membership function. Consequently, each variable is included in the membership function according to its spread among the population but independently of the deprivations observed for individual i for other commodities. For other indicators X_2, X_3 ... the procedure is similar. The degree of membership thus derives from the extent in which each of the poverty criteria are missing but this is done separately for each indicator. The main interest in the study of Cerioli and Zani (1990) is to open the way to the multidimensional relative measurement of poverty in a fuzzy context. Since the concept of poverty makes sense only in a given social context, this study is of major interest. Nevertheless, such a measure raises the issue of multicollinearity between different non-monetary indicators and between most of those indicators and the income itself. The main drawback of the Cerioli and Zani (1990) measure is that it does not permit the removal of the collinearity between two or more variables taken as components of the poverty measure. But, if there is a correlation between those variables, this may lead to an overestimated degree of membership. Let us take, for instance, the extreme situation where two variables are perfectly correlated. We then have redundancy using those two variables: they probably designate the same dimension of poverty. The weight of this dimension is thus twice what it should be. As a result, the Cerioli and Zani measure demands caution in choosing the variables belonging to the poverty membership function.

Two solutions may be envisaged to solve the correlation problem. The first one requires preliminary data analysis to avoid collinearity problems. The second way is to build a composite membership function in order to lower the weight of correlated variables. This allows keeping all the variables in the analysis and that is the new proposed method herein.

11.3.2 The proposed membership function

Let N be a set of n individual i (i=l, ..., n) and K a set of k monetary indicators j (j=l, ..., k). Each of the k indicators j takes their values in the interval [0,1].

The membership function is defined, for each individual i, according to the number of individuals having, at least, the same deprivations on each of the k indicators j. The higher the number of individuals with, at least, the same deprivations as individual i, the smaller the value of the membership function of individual i; that is to say the smaller the degree of poverty.

Let the proportion of individuals whose life style is less than or equal to the life style of individual i. The value of f_j is the number of individuals who are, at most, in the same position as individual i according to all the indicators. One must note that the proposed membership function of individual i relies only on this proportion f_j.

In order to build the membership function, we use a two-step procedure. First of all, use a first level membership measure for the poor set $m_P(i)$:

$$m_P(i) = \frac{\ln\left(\frac{1}{f_i}\right)}{\sum_{1}^{n}\ln\left(\frac{1}{f_i}\right)} \quad if \ 0 < f_i \leq 1 \tag{11.7}$$

The way f_j is built, the value of f_s is never equal to 0 since there is always, at least, one individual who has exactly the same deprivation as individual i: this is individual i him/herself. The second step leads to the proposed membership function:

$$\mu_P(i) = \frac{m_P(i) - Min[m_P(i)]}{Max[m_P(i)] - Min[m_P(i)]} \tag{11.8}$$

The degree of poverty is equal to 0 for all individuals who are in such a position that none of the other individuals has a better lifestyle. The poverty hazard increases with the value of the membership function. The poorest individual of the population has a value equal to 1 for the membership function. We consider that the degree of poverty is equal to one (fully poor), when individual i is in the least favourable position according to all the criteria taken simultaneously. This measure allows dealing with the collinearity issue since all the indicators are used jointly.

11.3.3 Example: calculation of a composite membership function

Let us assume 6 individuals and 3 dichotomous indicators. The latter are whether or not an individual has a bathroom, a car or goes to the cinema (this particular one is taken as a cultural indicator). X_1=1 if bathroom is present and X_1=0 otherwise. X_2=1 if a car is owned and X_2=0 otherwise. X_3=1 if an individual never goes to the cinema and X_3=0 otherwise. It is also assumed that the six individuals of this population are such as reported in Table 11.1.

Table 11.1 Calculation of a composite membership function

	X_1	X_2	X_3	f_i	$\mu_P(i)$
Individual 1	0	1	1	4/6	0.2
individual 2	1	1	1	1/6	1
individual 3	0	1	1	4/6	0.2
individual 4	0	0	0	6/6	0
individual 5	0	1	1	4/6	0.2
individual 6	1	0	0	2/6	0.6

By building, the degree of membership moves between the values 0 and 1. On one hand, the individual who has the highest deprivation (1,1,1) is allowed a degree of membership equal to 1. The extreme value of the scale concerns the individual who is in the best situation (0,0,0), because he holds the best life style. Between the two extreme values, one meets all the individuals who have only partial deprivation. One may observe, through this example, that going or not to the cinema (X_3) is correlated to the possession of a car (X_2). Moreover, going to the cinema or being the owner of a car doesn't constitute the standard of life because 2/3 of individuals are deprived of these two variables. Consequently, people who are only deprived of these 2 variables are not really to be considered as poor and then their degree of membership is rather low (0.2). On the other hand, the individual 6 who has only one deprivation (no bathroom) has higher degree of membership (0.6) because of the bigger spreading of the bathroom among the population, even if he has only has 1 deprivation.

The range of the degrees of membership is by construction from 0 to 1. An individual, who has the least favourable position, has a degree of membership equal to 1. At the other extreme of the scale, one finds the individual who has the best life style since he/she profits from all the pleasures selected here ($\mu_P(i)=0$). Between these two extremes we find the group of those revealing only partial lacks. One observes, through this example that the cinema admissions (X_3) are correlated with the possession of a car (X_2), and that in addition, being the owner of a car, just like going to the cinema does not form part of the dominating way of life, since two thirds of the individuals are deprived. Thus, the individuals, without these two pleasures, are exposed little to poverty; their degree of membership in the group of the poor is thus relatively low ($\mu_P(i)=0.2$). On the other hand, individual 6 for whom one observes only one deprivation (bathroom) is associated with a stronger degree of poverty ($\mu_P(i)=0.6$), because the equipment in the bathroom forms part of the dominating way of life.

11.4 Empirical comparison on French Youth Panel Survey from 1996 to 1999

In this section we shall present an empirical comparison of the three concepts of poverty based on the fuzzy method proposed in Sect. 11.3 and we shall see whether the choice in favour of certain normative principles has consequences on the population identified more or less as being poor. After presenting some preliminaries, this section considers the way in which the concepts must be interpreted for young people undergoing transition from school to work. Finally, we compare the results based on primary goods with those obtained using other concepts for measuring poverty.

11.4.1 Preliminaries

In order to analyse poverty, three considerations are imperative. Firstly, such an application was conducted by the third panel of the French Centre of Research in Education, Training and Employment (CEREQ). People who left school in 1994 with at the most the "baccalaureate" degree are surveyed respectively from October 1993 (beginning of their last academic year) to February 2000. This cohort may exist from regular secondary education, Specialised Instruction Section (SES) and apprenticeship. Five annual series of interrogation were carried out between 1996 and 2000. The database we used consists of 2297 individuals for 1996 to 1999 (see Appendix 1).

Secondly, how can we measure young people's poverty that would be based on primary goods, basic capabilities and primary outcomes? In order to remain faithful to Rawls and Fleurbaey the ideal solution would probably be to conduct an empirical study on the lists proposed in Sect. 11.1. Quite apart from the inherent limitations of the process (a constitutional state assures the same rights and liberty for everyone, the social bases of self-respect is rather difficult to measure...), the significance of such an application is clearly questionable because of the specificity of the population studied. Consequently, we suggest, as Sen does, to take time to reflect on what is relevant for young people in their general attitude to each of the three concepts. Let us emphasize that principles are very theoretical and poverty is impossible to measure strictly with respect to the definitions. There are of course many difficulties with theoretical concepts. In particular the problem of applying primary goods is a serious one. This is the occasion to go into operational arbitration in the concepts. In consequence, we have tried to translate empirically the theoretical concepts in the specific context of French school leavers undergoing a period of transition from school to work. But the empirical analysis is very tentative and is certainly explanatory. Indeed, apart from this work, we don't know of any empirical analysis based on the concepts used and applied in this field.

Thirdly, as mentioned in Sect. 11.1, it is impossible from the point of view of the application to take support from capabilities, because the data generally provides information on the actions and the states reached rather than on the whole of the actions and the states to the range of the individuals. To approach the concept of capability, we thus followed Sen (1992), who proposes considering refined functionings.

11.4.2 The informational basis of primary goods

In the database of the panel of CEREQ, we identified a certain number of indicators of resources and classified them in three categories, housing conditions, education, and wealth. A short illustration of the characteristics of each indicator follows; all of the variables which we considered are rather heterogeneous, in order to cover several aspects of the means of achievement of functionings (see Table 11.2).

Education is considered here as a means of achieving various functionings. It is thus supposed to have its own function in determining capabilities and functionings. It is true that the source used presents extremely complete information on the received initial formation. It could thus be completely convenient to work on the basis of a much finer variable, in particular to study the relation between the formation, under consideration as means of insertion, and the position on the labour market, intended as functioning of professional insertion. But such an analysis exceeds the framework of this study and we choose to concentrate on three elementary indicators. In the subset of variables we integrated education, the initial training level, information relating to the achievement of diplomas and the characteristic associated with the possession of a driving licence. This sum of money element is collected in the subset of the focal variables of formation, because in our view, it constitutes a human capital which influences with whole share capabilities and functionings. Three training levels are distinguished.

Table 11.2. Deprivations on "primary goods"

	Women	Men	Total
Education			
• Level IV	29.9	21.7	26.8
• Level V	45.9	52.0	49.7
• Level Vb and VI	32.4	18.1	23.5
No diploma at the end of schooling	33.6	39.8	37.4
Without any driving licence	19.6	8.8	12.9
Housing equipment			
Neither shower nor bath	0.4	0.5	0.4
No water closed	1.1	1.1	1.1
No warm water	0.2	0.3	0.2
No heating	0.3	0.2	0.3
No furnace	3.2	3.0	3.1
No form of refrigeration	0.3	1.2	0.8
Monetary resources			
No saving of money	72.1	78.7	74.6
No family help	66.3	67.9	66.9
Income from activity			
• < 2160 F	23.6	10.7	15.6
• ∈ [2160 F, 3480 F]	17.4	9.0	12.2
• > 3480 F	58.9	80.3	72.1

CEREQ panel data survey - Wave 4
Reading: 19.6% of the young women and 8.8% of the young men do not have a driving licence

The individual income of activity is not a very widespread concept. It concerns any form of resource related to a present activity, such as wages, training allowance, or even to a former activity, such as unemployment allowance. More precisely, information used together with the gross monthly income of activity at the date of investigation. It amounts on average to 4786 F, as one can read in Table 11.3. This value is naturally below the level of gross monthly wages.

The income from activity appears among the class of the variables of the continuous type. The limit of the two borders is fixed at 40% and 60% of the median income of activity, that is to say 2160 F and 3480 F, respectively.

Table 11.3. Individual income from activity (in Francs)

	Women	Men	Total
Average of activity income Me- dian of activity Income	3854 4000	5362 5800	4786 5400

CEREQ panel data survey - Wave 4

11.4.3 The informational basis of primary social outcomes

In the panel of CEREQ, some indicators of outcomes have been identified as various things that one accomplished during his life. Objects of value are classified in four distinct categories: residential independence, leisure activities, debt and finally situation on the labour market. Firstly, information on residential independence was mobilised, because in our view, the achievement of this functioning has constituted a significant stage in the current context of France for young people who have completed their school course for five years and entered adulthood. Secondly, three kinds of deprivation were listed according to leisure: holidays, outings and time devoted to domestic tasks. When the time spent on these types of activities exceeds ten hours a week, a deprivation is noted on the level of this characteristic. Thirdly, information on the debt is included. It reflects the facility with which an individual succeeds in ensuring the management of his budget; this variable was built on the basis of a particular question of the survey. Thus, individuals who stated that they had been involved in debt are regarded as having a deprivation. Fourthly, outcomes relate to the position occupied on the labour market at the date of investigation. The situations are distinguished according to three categories: employment, unemployment and other situations (training or taking up studies again, or of national service or finally of inactivity. Here, there are categorical variables presenting more than two alternatives.

Table 11.4. Deprivation on outcomes (%)

	Women	Men	Total
Labour market position			
• Employment	67.1	79.3	74.7
• Unemployment	24.1	15.8	18.9
• Other situation	8.2	4.9	6.4
Leisure			
Never goes on holiday	38.6	36.3	37.2
Never goes to the cinema, the theatre, concert, etc.	19.1	9.7	13.3
Spent more than 10 hours per week on domestic tasks (kitchen, household)	42.9	14.9	25.6
Independence	34.0	59.9	50.0
Live with his/her parents			
Debt	6.4	6.0	6.1
Spent more than he/she earns and is involved in debt			

CEREQ panel data survey - Wave 4.
Reading: 14.9% of the men devote more than 10 hours per week to domestic tasks

11.4.4 The informational basis of refined functionings

In agreement with what has been explained in Sect. 11.1, we shall try to work on refined functionings instead of capabilities. It is in fact only possible to characterize functionings in a "refined" way in order to take note of the counterfactual opportunities. Corresponding to the functioning x, a "refined functioning" takes the form of "having functioning x through choosing it from the set S". It was possible to turn only two functionings into refined functioning: firstly, residential independence; and secondly position on the labour market.

Firstly, as far as housing independence is concerned we had recourse to one second question present in the survey. It aims at including understanding the reasons for which the young people questioned remain at home with their parents. This question does not tell us anything about real opportunities of choice of young people no longer living under the same roof as their parents. In this context, we thus decided to distinguish two situations respectively classifying opportunities: "to live in their parents' house from need" and "to live in their parents' house from choice". We considered young people living in their parents' house from need, if they claimed economic reasons for doing so or if they claimed to be waiting for employment on a permanent contract. In all the other situations we thus admitted that they lived with their parents by choice. It will thus be considered that there is a deprivation from the point of view of refined operations, since the decision to remain with his/her parents is not deliberated.

Secondly, as far as the labour market position is concerned, we can make use of additional questions about possible alternatives. Finally, one is in the presence of multiple situations differentiated according to opportunities from choice, which we classified as most favourable to most constraining in the following way:

Employment or training or taking up studies or inactivity from choice
Unemployment with employment proposals
Employment from need
Unemployment with no employment proposals or inactive with resignation or national service.

One was thus brought to establish a categorical variable collecting four distinct modalities for which a degree of membership was built in reference to the method presented.

Table 11.5. Deprivation on refined functioning (%)

	Women	Men	Total
Labour market position			
• Employment or training or taking up studies again or inactivity from choice	67.5	78.1	74.1
• Unemployment but had refused some employment proposals	2.6	1.8	2.1
• Employment from necessity	2.9	3.2	3.1
• Unemployment and never had any employment proposal	27.0	17.0	20.8
Leisure			
Never goes on holiday	38.6	36.3	37.2
Never goes to the cinema, the theatre, concert, etc.	19.1	9.7	13.3
Spent more than 10 hours per week on domestic tasks (kitchen, household)	42.9	14.9	25.6
Independence			
Live with his/her parents	25.9	48.4	39.8
Debt			
Spent more than he/she earns and is involved in debt	6.4	6.0	6.1

CEREQ panel data survey - Wave 4

11.4.5 Analyse recovery of the three populations

It is extremely interesting to compare the results based on primary goods with the ones obtained using other concepts to measure poverty. As mentioned before, three membership functions are drawn up to rank individuals based on the primary goods, outcomes and refined functionings criteria. As would be expected, the ranking of the three subgroups differs slightly depending on which concept was chosen. When the same percentage of poorest individuals (approximately 10%) is isolated at the bottom of three membership degrees of poverty, three populations are found with only partial coverage. Whereas more than 20% of the population bears at least one of the marks of poverty, only 1,7% bear the three at the same time. Figure 11.2 illustrates the various situations.

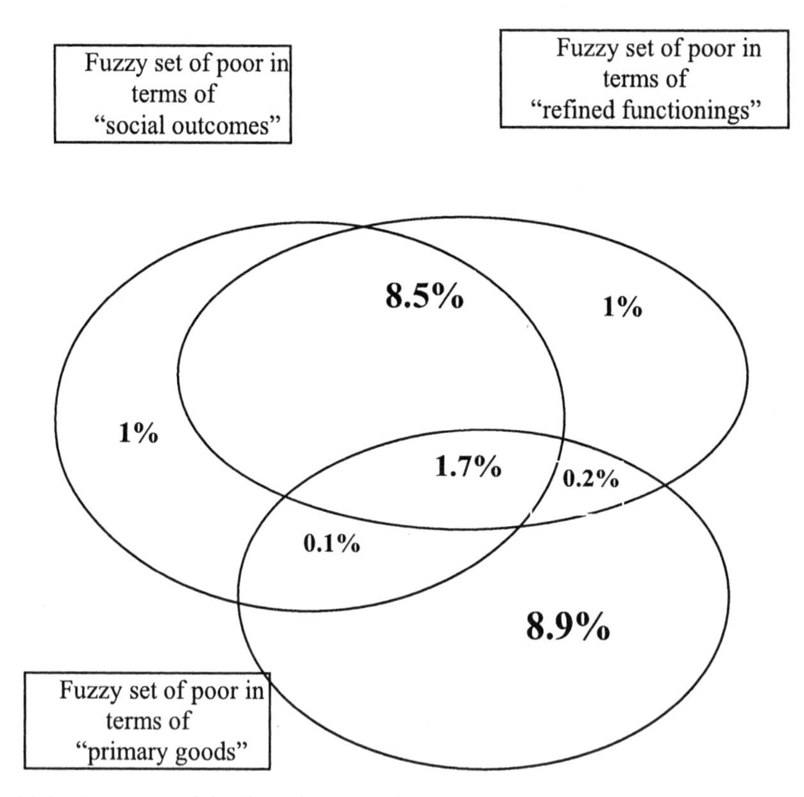

Fig. 11.2. Recovery of the three images of poverty

The results present in Figure 11.2 are now analysed in more detail. The ranking provided by outcomes analysis is surely closer to the ranking provided by the refined functionings approach. Indeed, 90% of individuals whose intensity of poverty is the highest according to outcomes also have the highest level of poverty according to refined functioning and vice versa.

Finally when poverty of individuals is estimated using outcomes or refined functionings, more or less the same image of poverty is obtained. Poverty according to outcomes thus tends to be combined with poverty according to refined functionings, without mistaking one for the other. This is probably due to the way functionings are refined. Only two functionings have been refined. But as already mentioned, there was not enough information for selecting a different set of functionings other than outcomes in each case.

As a matter of fact the ranking provided by the primary goods concept is completely different from the other rankings. The poorest according to primary goods are for a very large proportion, not the same people as the ones obtained using other concepts to measure poverty. Indeed, 75% of individuals whose intensity of poverty is the highest according to outcomes, don't have the highest level of poverty regarding primary goods. According to refined functioning, the difference in the ranking of poorest people is very important when primary goods evaluate poverty. The differences are statistically significant.

11.5 Conclusion

In this conclusion we propose to return to the essential features of our work. Our initial motivation proceeded an examination of the question of the recovery between three forms of poverty. The concept of poverty was considered under three different ethical styles privileging first of all primary goods, secondly social outcomes and lastly basic capabilities. The most important finding to emerge from our research is that the use of a specific concept of poverty would alter the ranking of people in a poverty scale: It has been particularly confirmed when one compares primary goods with social outcomes or functionings. Therefore one must first choose the objects of value in accordance with the value judgments involved (Vero 2004). It means that one is forced to ask over which kind of variable individuals must have control and for what sort of variable society is responsible. So the first relevant question for measuring poverty is, as Sen mentioned: "Equality of what?" This question is likely to return to very pressing problems about such things as real interests. But of course this is an open question.

Appendix 1 - The CEREQ Panel Data Surveys

The French Centre for Research on Education, Training and Employment (CEREQ) in collaboration with the Department of research and statistical survey (Dares) of the ministry of Employment and Solidarity carried out a third panel of "youth measures" among a sample of 3500 young people who had left school in 1994 with initial education lower than or equivalent to the baccalaureate. The main purpose of this survey was to provide data on the use of youth programs to ease the school to work transition. The sampling frame was based on lists of former pupils gathered from secondary schools (*lycées* and *colleges*) and on apprenticeship contracts supplied by the Ministry of Employment and Solidarity. The panel survey comprised five annual waves and was performed using the Computer Assisted Interview Procedure (CATI). The themes broached during the interviews concerned initial education, occupational pathways (month-by-month progress report after leaving the educational system in order to avoid memory bias), family background, income and living conditions.

Table 11. 6. Attrition rate

Frequency	No. of respondents	Attrition rate (%)
Wave 1	3469	
Wave 2	2957	15
Wave 3	2627	11
Wave 4	2297	13

Appendix 2 – French Educational Level

Level IV: Leaving last class secondary education: general "baccalaureate", technological "baccalaureate", vocational "baccalaureate", and Technician's certificate.

Level V: Leaving upper secondary education before last class ("terminale"), or last year of first level vocational preparation: third year of three year CAP, second year of two year BEP, second year of two year CAP, supplementary certificate to CAP or BEP.

Level Vbis: Leaving lower secondary education or first level vocational preparation before the last year: first year of three year CAP, second year of three year CAP, first year of BEP, first year of two year CAP.

Level VI: Early leaving (from 6th, 5th or 4th) or pre-vocational preparation: primary studies certificate (CEP), Pre-vocational Class (CPPN), Preparation for Apprenticeship (CPA).

References

Cerioli A, Zani S (1990) A Fuzzy Approach to The Measurement of Poverty. In: Dagum C, Zenga M (eds) Income and Wealth Distribution, Inequality and Poverty. Springer Verlag, Berlin, pp 272-284

Cheli B, Ghellini G, Lemmi A, Pannuzi N (1994) Measuring Poverty in the Countries in Transition via TFR Method: The case of Poland in 1990-1991. Statistics in Transition 5:585-636

Cheli B, Lemmi A (1995) A "Totally" Fuzzy and Relative Approach to the Multidimensional Analysis of Poverty. Economic Notes 24:115-134

Dubois D, Prade H (1980) Fuzzy Sets and Systems Theory and Applications. Academic Press

Fleurbaey M (1995) Equal Opportunity or Equal Social Outcome ? Economics and Philosophy 11:25-55

Fleurbaey M (1998) Equality among responsible individuals. In: Laslier JF, Fleurbaey M, Gravel N, Trannoy A (eds) Freedom in Economics. Routledge, London, pp 206-234

Muellbauer J (1987) Professor Sen on the Standard of Living. In: Hawthorn G (ed) The Standard of Living, pp 39-58

Rawls J (1971) A Theory of Justice. Harvard University Press

Sen AK (1980) Equality of What ? In: McMurrin SM (ed) Tanner Lectures on Human Values. Cambridge University Press

Sen AK (1985) Commodities and Capabilities. North Holland

Sen AK (1987a) The Standard of living. Cambridge University Press

Sen AK (1987b) On Ethics and Economics. Oxford Blackwell Publishers

Sen AK (1992) Inequality Reexamined. Clarendon Press

Vero J, Werquin P (1997) Un réexamen de la mesure de la pauvreté. Comment s'en sortent les jeunes en phase d'insertion professionnelle ? Economie et Statistique 308-309-310:143-148

Vero J (2002) Mesurer la pauvreté à partir des concepts de biens premiers de réalisations primaires et de capabilités de base. Le rôle de l'espace d'information dans l'identification de la pauvreté des jeunes en phase d'insertion professionnelle. Ph.D. thesis, EHESS

Vero J (2003) A la recherche d'un concept de pauvreté: les théories économiques de la justice en héritage. Revue de l'Economie Méridionale 201-202:35-45

Zadeh LA (1965) Fuzzy Sets. Information and Control 8:338-352

Zadeh LA (1970) Decision making in a fuzzy environment. Management Science 17:141-164

12 Multidimensional Fuzzy Relative Poverty Dynamic Measures in Poland

Tomasz Panek

Institute of the Statistics and Demography, Warsaw School of Economics

12.1 Introduction

The pro-market reforms launched in Poland in 1989 and their continuation over the most recent decade should generally be acknowledged as a success. After an initial recession the Polish economy embarked on a path of steady growth (WSE 2000). 1999 was the 6th year in a row that the GDP was on the rise with a growth rate exceeding 4%. During the most recent decade domestic prices[1] and public finances were stabilized, most state-owned enterprises were privatized.

The social costs of transformation, however, proved to be very high. Although the vast majority of the society supports the market system, many social groups believe that the economic and social costs of transformation were excessive and unjustly distributed. The growth in the unequal distribution of income and growing unemployment[2] meant that not all groups in the population took advantage of economic growth to the same degree. For many of them the result of economic growth was a reduction in their absolute standard of living or at least relatively, i.e. in comparison with other groups of the population.

After the first years of reduction in the population's standard of living, the real household income has been growing regularly. Nevertheless, the high prices of durables, especially apartments, caused substantial disparity between the household income earned and the non-income aspects of material living conditions. The complexity of the processes determining the material conditions of households shows that analyses of the changes in the poverty sphere in Poland should go beyond the income factor. One should also incorporate other dimensions of material living conditions, primarily associated with an evaluation of household affluence and hous-

---

[1] Inflation in 1990 amounted to 30-40% per month while by 1999 it fell to 7.3% per annum and continued to show a downward trend.

[2] The unemployment rate at the end of 1999 was 13.1%, while at the end of 2000 it stood at 15% and continued to show an upward trend.

ing conditions. The multi-dimensional approach will therefore have particular application to poverty analyses.

The fundamental objective of this research was to analyze the changes in the degree of the poverty threat, treated multi-dimensionally, in Poland from 1996 to 1999 in the basic socio-economic groups of households. Moreover, an assessment of the nature of poverty in Poland was made in its individual dimensions, by distinguishing between transitory and chronic poverty. Finally, probit models were used to specify the impact exerted by the households attributes on the degree of the poverty threat.

12.2 Sources of Data

The basis for the analyses conducted are the data from the household budget survey conducted by the Polish Central Statistical Office from 1996 to 1999 (PCSO 1999). The panel approach was applied in this research whereby the same households were observed in consecutive years. To preserve the panel nature of the sample, information concerning only 4,485 households was used as these households were observed in all of the consecutive years from 1996 to 1999. In order to maintain the representative nature and the initial structure of the sample in all phases of the panel research, the proper weighting systems were applied. The weightings depended on the frequency with which households withdrew from the panel in the classification cross-sections by socio-economic groups of households.

Selected information provided by the household budget research has been used in this research. This includes basic demographic and socio-economic characteristics of households and their members, information about cash and non-cash income and the amount of expenditures broken down by their main groups, housing conditions, selected durable goods and subjective assessments made by households of their overall material standing.

Household income and expenditures in the individual months of a given year are expressed in average monthly prices of a given year. The category of income applied in the research was the net monthly disposable household income. In the research classification cross-sections of households by socio-economic group (source of maintenance) was used.

12.3 Methods of Analysis

12.3.1 Multidimensional Analysis of Poverty

Identification of the Poverty Sphere

The multidimensional analysis of poverty has used a method based on the fuzzy set theory (Zadeh 1965; Dubois and Prade 1980). It allows one to avoid designating the boundary (poverty line) sharply dividing all households under examination into poor and non-poor. In addition to poor and non-poor households an additional intermediate group of households with a varying degree of poverty threat has been distinguished.

The fuzzy set theory was applied in practice, for the first time, in the sphere of poverty research in Italy (Cerioli and Zani 1990). Its adaptation to Polish conditions was conducted by Błaszczak-Przybycińska (1992) and Panek (1994)[3].

By using the fuzzy set theory it is possible to determine the degree to which every i-th household belongs to the sub-set of impoverished households, defining the poverty membership function.

The appropriate definition of the function of belonging to the sub-set of impoverished households (membership function) for each of the variables characterizing the distinguished dimensions of poverty is therefore crucial in this approach to the multidimensional analysis of poverty. Membership function specifies the extent and intensity of poverty jointly.

In poverty analysis two complementary approaches have been applied in parallel: the objective approach and the subjective approach (Panek et al. 1999). The terms "objective" and "subjective" should not be associated with the degree of arbitrariness applied when measuring poverty. In each one of these means of measurement, there are certain arbitrary findings. In the objective approach the evaluation of the degree to which the needs of the households under examination are satisfied is carried out notwithstanding their personal valuations. In the subjective approach the households under examination evaluate themselves the degree to which their needs have been satisfied. In the objective approach when measuring the degree of the poverty threat posed to households, housing conditions and household affluence have been taken into consideration in addition to their in-

---

[3] Compare also the paper by Cheli et al. (1994).

come standing. This constitutes a certain limitation on the dimensions of poverty resulting nevertheless solely from the scope of the available data[4].

Specification of the Poverty Membership Function

The starting point for multidimensional analysis of poverty is the specification of these attributes and the behavior of households whose specified variants (or quantitative figures) may be treated as symptoms (indications) of poverty. In the objective approach these symptoms, for instance, were as follows: the fact that households did not own certain durable goods (in the group of needs entitled *household affluence*), income level too low (in the group of needs entitled *income*) or excessive density of housing held by households (in the group of needs entitled *housing conditions*). In the subjective approach the occurrence of poverty symptoms was specified on the basis of the evaluation of a household's general material standing made by the household itself. This evaluation was made on the basis of the household choosing one of five variants (from very good and rather good to average to rather bad and bad), i.e. the one most suitable to the household's standing.

The determination of the degree of membership of each household in the poverty sphere by the distinguished symptoms entails the transformation of the empirical values of the households' poverty symptoms into values defined as the degree of membership in the poverty sphere. This transformation is carried out through the membership function. The household's poverty membership function due to a non-dichotomous variable is described by the following: $\mu(x_{ij_d d}) = 1$ when the j-th variable, from the d-th dimension of poverty indicate the existence of a poverty symptom in the i–th household; $\mu(x_{ij_d d}) = 0$ in the opposite situation; where: i=1,2,...,n, j=1,2,...,k_d, d=1,2,...,l.

As introduced in Chapter 3, if a variable, e.g. disposable income, is non-dichotomous Cerioli and Zani (1990) considered a transition zone $x_j^L < x_{ij} \le x_{ij}^H$ for attribute j over which the membership function declines from 1 to 0 linearly:

[4] Broader-based inclusion of the dimensions of poverty were applied in the research into household living conditions conducted by the research team at the Warsaw School of Economics (Panek et al. 1999).

$$\mu_j\left(x_{ij}\right)=\begin{cases}1 & if \quad x_{ij}\le x_j^{(L)}\\ \dfrac{x_j^{(H)}-x_{ij}}{x_j^{(H)}-x_j^{(L)}} & if \quad x_{ij}\in\left(x_j^{(L)},x_j^{(H)}\right]\\ 0 & if \quad x_{ij}>x_j^{(H)}\end{cases} \tag{12.1}$$

Developing this solution Cheli et al. (1994) proposed the TFR method (*Totally Fuzzy and Relative*), in which the value of the membership function in the poverty sphere on account of the non-dichotomous variable is estimated on the basis of the empirical distribution function designated after putting into order the empirical values of this variable according to the increasing poverty threat. We may write this function as $\mu(x_{ij})=1-F(x_{ij_d})$, where F is the empirical distribution function of the j-th variable, from the d-th dimension of poverty.

This solution based on the relative approach to poverty measurement (Panek et al. 1999), makes its assessment totally dependent on the degree of inequality in the distribution of the needs satisfaction level (described by symptoms of poverty), and not by the needs satisfaction level of the household itself. In our opinion, the evaluation of inequality in the needs satisfaction level of households should be based on inequality indices calculated separately and not decided entirely on the degree of the poverty threat (values of membership function to the poverty sphere in our case). We propose a solution in which established normative threshold values would decide on the total membership in the poverty sphere and the total non-membership in the poverty sphere. In turn, the degree of membership in the poverty sphere when the empirical value of the poverty symptom (within the framework of a given dimension of poverty) is between the threshold values, depends on the empirical distribution of this symptom in the population under examination.

Ultimately, the form of the household's membership function in the set of poor households by a non-dichotomous variable is described by the formula presented in Sect. 3.5 of Chapter 3:

$$\mu_j\left(x_{ij}\right)=\begin{cases}1 & if \quad x_{ij}=x_j^{(s)}\\ \mu_j\left(x_j^{(l-1)}\right)+\dfrac{F(x_j^{(l)})-F(x_j^{(l-1)})}{1-F(x_j^{(l)})} & if \quad x_{ij}=x_j^{(l)}\\ 0 & if \quad x_{ij}=x_j^{(1)}\end{cases} \tag{12.2}$$

The estimation of the membership function by a given non-dichotomous variable using formula (12.2) requires that its values obtained in individual households first be ordered by the increasing poverty threat. In turn, the estimate of the membership function by immeasurable non-dichotomous variables requires the prior transformation of these variables into measurable variables by assigning quantitative ranks to their variants.

In the analysis of the income dimension of poverty the assumption was made that the basic measure of a household's income is its current monthly disposable income divided by the equivalence scale calculated for it. The income estimated according to these assumptions is called equivalent disposable income. Equivalency scales are parameters which make it possible to compare the income of households with varying characteristics to the poverty boundary calculated for the household constituting the reference point (the standard household whose equivalency scale is 1). The equivalence scale for a given type of household shows how many times its income would have to increase (or decrease) to reach the same level of consumption equal to the level of need satisfaction in the standard household. The equivalence scales applied in the research were estimated on the basis of an estimate of econometric models. It used information concerning the amount of household expenditure, household demographic characteristics and consumer goods prices (Panek et al. 1999).

The thresholds for the variable disposable equivalent income were the adjusted social minimum and the adjusted subsistence minimum for a single-person household, as estimated by the Institute of Labor and Social Affairs[5]. The value of the social minimum and the subsistence minimum are identical to the values of the consumer goods baskets (Deniszczuk and Sajkiewicz 1996a, 1996b). The value of the basket for the social minimum should ensure such household living conditions as being able to reproduce its life forces as well as being able to have and rear children and maintain ties with society. This designates the minimum level of income ensuring a "dignified" way of life, and consumption suitable for the average level of welfare in the nation. In turn, the scope of the basket for the subsistence minimum designates the boundary below which there is a threat to the life and to the mental and physical development of a person.

The values of the membership function for households in the poverty set in distinguished dimensions of poverty as described by sets of more than one variable were calculated by using the arithmetic weighted average formula, which is:

[5] The principles of calculation of both minimums leads to ever greater revaluation in subsequent years of research, and thereby to their incomparability in dynamic analyses. The adjustment made to them has made them comparable in subsequent years of research (Panek 2001).

$$\mu(x_{id}) = \frac{\sum_{j=1}^{k_d} \mu(x_{ij}) w_j}{\sum_{j=1}^{k_d} w_j} \tag{12.3}$$

where $w_j = \ln \frac{1}{\overline{\mu}(x_j)}$ and $\overline{\mu}(x_j) = \frac{1}{n}\sum_{i=1}^{n} \mu(x_{ij})$.

The weighting system used makes the values for the individual symptoms of poverty (the variables) dependent on their frequency of occurrence within the population. The symptoms of poverty which occurred the least frequently were considered to be the most significant (they were given more weight) since the infrequency of their occurrence indicates that they pertain to the households' basic needs. At the same time, using a logarithmic function in constructing a system of weights makes it possible to avoid assigning excessive significance to poverty symptoms.

The membership function values in the poverty sphere of the i-th household have been estimated on the basis of equations (12.1) and (12.2). The membership function values in the poverty sphere of a given household by all distinguished dimensions of poverty in total have been calculated on the basis of the following formula:

$$\mu(x_i) = \frac{1}{l}\sum_{d=1}^{l} \mu(x_{id}) \tag{12.4}$$

In aggregating the membership function values of individual households in the poverty sphere (12.4) we derive the poverty threat index for the entire population under examination (Cerioli and Zani 1990):

$$P = \frac{1}{n}\sum_{i=1}^{n} \mu(x_i) \tag{12.5}$$

12.3.2 Evaluation of the Poverty Nature

In dynamic analyses of the poverty phenomenon it is extraordinarily important to determine whether a household was in the sphere of a strong poverty threat temporarily or whether the threat is chronic (Rodgers and Rodgers 1993; Betti 1996; Stevens 1999). This is of exceptionally important significance when articulating social policy projects to fight poverty. They should concentrate on counteracting chronic poverty.

The evaluation of the degree to which the strong poverty threat is chronic in Poland was based on an analysis of the number of years during which households are in the sphere of the highest poverty (households for which the membership function is no less than 0.50 in the objective approach and households which evaluated their material situation as bad or rather bad in the subjective approach) and on transition matrices describing the mobility of households in terms of their membership to the distinguished classes of the degree of the poverty threat (classes of membership function values) in subsequent years of the research.

Moreover, mobility measures were used to analyze household mobility based on the Shorrocks measure of mobility (Shorrocks 1978), which is described by the formula:

$$S = \frac{n - tr(P)}{n-1} \tag{12.6}$$

where tr(P) is the trace of the transition matrix, and $P=[n_{ij}]$ is a transition matrix.

The measure (12.6) assumes values from the range $\left[0, \frac{n}{n-1}\right]$. The higher the measure value the greater the mobility of the households. When normalizing the measure (12.6), so that it always assume values in the range [0;1] and when decomposing it, we ultimately derive the following:

$$S = \frac{n - tr(P)}{n-1} = \frac{\sum_{i>j} n_{ij} + \sum_{i<j} n_{ij}}{n-1} = \frac{\sum_{i>j} n_{ij}}{n-1} + \frac{\sum_{i<j} n_{ij}}{n-1} = SU^{+} + SU^{-} \tag{12.7}$$

The first of the elements on the right side of the equation points to the percentage of households featuring a decline in the degree of the poverty threat in the periods under comparison. The second element in this equation is the percentage of households in which this threat increased during the period under research. The measure of the nature of household mobility was also estimated as a complement to the mobility measure:

$$CM = \frac{\sum_{i>j} n_{ij}}{n-1} - \frac{\sum_{i<j} n_{ij}}{n-1} = SU^{+} - SU^{-} \tag{12.8}$$

This measure assumes values in the range [-1;1]. Positive values indicate more household flows from groups with a higher poverty threat to groups with a lower poverty threat. Negative measure values show more flows increasing the poverty threat than flows reducing the poverty threat. The higher the absolute value of the measure the greater the advantage of one type of flow over the other type of flow.

12.3.3 Poverty Determinants

One method of analyzing the sources of poverty frequently used in practice is the division of the population under research into groups according to selected socio-economic traits and the evaluation of this phenomenon with the use of poverty indices. High poverty indices in a given household group, coupled with a high variation among groups according to a given classification, suggests that a given variant of a trait characterizing the distinguished group of households generates poverty.

Evaluations of the impact exerted by individual variables on generating poverty in isolation may prove to be burdened since they do not take into consideration the linkage between these variables and other variables. For example, strong poverty threat in the group of rural households suggests that residing in a rural area generates poverty. However, strong poverty threat in this group of households is the combined effect not only of rural residence but also of other factors. Therefore in order to specify the determinants of poverty it is necessary to estimate the net impact of individual variables on the generation of poverty, which requires the application of multidimensional methods of researching co-dependencies, especially multidimensional regression. To specify the impact exerted by the attributes distinguished in the research on the degree of the poverty threat, probit models (Greene 1997) were applied to poverty analysis using the objective or subjective approach. In these models the dependent variable is the dummy variable which assumes a value of 1 when a household was in a higher poverty sphere (in the objective approach when the membership function value was no less than 0.75, and in the subjective approach when households evaluated their material situation as rather bad or bad) through all the years of the research and 0 otherwise.

The model used in the research may be written as follows:

$$\Phi^{-1}[p(x)] = \alpha_0 + \alpha_1 x_1 + \alpha_2 x_2 + \ldots + \alpha_k x_k + \varepsilon \tag{12.9}$$

Where $\mathbf{x}$ is a vector of prospective determinants of poverty (explanatory variables), p($\mathbf{x}$) is the risk of a household being in strong poverty threat for the entire research period against a specified potential arrangement of the determinants of poverty (independent variables), Φ^{-1}(p) is the inverse function of the standard normal distribution function, and ε is the model's random error.

The explanatory variables incorporated in the models as possible determinants of poverty have been presented just like the explained variable using arrangements of dummy variables. When estimating models with arrangements of dummy variables, in each one of these arrangements, one of the binary variables (variants of the attribute) was bypassed in order to avoid co-linearity. This means that the parameters next to the model's independent variables are relative measures of the risk of entering the sphere of high poverty risk. Explanatory variables in both approaches were the basic socio-economic characteristics of households. Furthermore, the membership function values in the poverty sphere in terms of the income situation, housing conditions and household affluence were considered as possible determinants of a high risk of poverty in the subjective approach.

12.4 Changes in the Poverty Sphere in Poland from 1996 to 1999

12.4.1 Degree of the Poverty Threat

The mean membership function values were used as the basis to evaluate the changes to the poverty threat in Poland from 1996 to 1999. The membership function values in the poverty sphere in the objective approach have shown a downward trend in all the distinguished dimensions in the period under examination (Table 12.1). The poverty threat in all the years under examination was clearly the highest in the household affluence dimension. In turn, we have observed the lowest poverty threat in 1996, which was the first year of the analyses, in housing conditions. Nevertheless, in subsequent years of research the housing conditions of the households, like household affluence, improved much more slowly than household income standing. This is shown by the changes to the membership function values in the poverty sphere for the non-monetary dimension (housing conditions and household affluence treated jointly). This is understandable since poverty symptoms are resources in nature for the non-monetary dimension of poverty, and thus subject to gradual changes in most households. In terms of household income, which is a stream in nature, the degree of the poverty threat was subject to much greater volatility over the years of research. These regularities have meant that since 1997 the lowest poverty threat to households in the objective approach was present in the income dimension. The poverty threat in the entire period under analysis was much higher according to households' evaluations (subjective approach) than indicated by the measurement in the objective approach (Table 12.1). The causes of this situation should be seen in the fact that household aspirations concerning their material standing substantially surpass the standards accepted at the current level of the nation's socio-economic development.

Table 12.1. Poverty Threat by Socio-economic Group in 1996-1999

Socio-economic group	Average values of membership function					
	objective approach					subjective approach
	income	housing conditions	household affluence	non-monetary	total	
Employees						
1996	0.101	0.103	0.181	0.142	0.128	0.509
1997	0.089	0.099	0.167	0.133	0.118	0.513
1998	0.074	0.093	0.152	0.122	0.106	0.515
1999	0.058	0.085	0.147	0.116	0.097	0.507
Employee-farmers						
1996	0.213	0.178	0.250	0.214	0.214	0.490
1997	0.160	0.181	0.223	0.202	0.188	0.517
1998	0.134	0.169	0.211	0.190	0.171	0.522
1999	0.121	0.160	0.206	0.183	0.162	0.532
Farmers						
1996	0.459	0.250	0.307	0.279	0.339	0.496
1997	0.353	0.227	0.280	0.254	0.287	0.501
1998	0.341	0.198	0.255	0.227	0.265	0.536
1999	0.352	0.184	0.244	0.241	0.260	0.578
Self-employed persons						
1996	0.086	0.071	0.110	0.090	0.089	0.317
1997	0.065	0.075	0.111	0.093	0.084	0.334
1998	0.072	0.069	0.091	0.080	0.077	0.351
1999	0.069	0.065	0.083	0.074	0.072	0.341
Retirees and pensioners						
1996	0.142	0.142	0.321	0.232	0.202	0.526
1997	0.100	0.147	0.312	0.230	0.186	0.542
1998	0.090	0.135	0.292	0.213	0.172	0.549
1999	0.089	0.134	0.295	0.214	0.172	0.555
retirees						
1996	-	-	-	-	-	-
1997	-	-	-	-	-	-
1998	0.046	0.121	0.284	0.202	0.150	0.489
1999	0.043	0.115	0.288	0.202	0.149	0.497
pensioners						
1996	-	-	-	-	-	-
1997	-	-	-	-	-	-
1998	0.165	0.159	0.303	0.231	0.209	0.648
1999	0.197	0.165	0.306	0.235	0.212	0.655
Persons living on unearned sources						
1996	0.513	0.197	0.325	0.261	0.345	0.798
1997	0.535	0.212	0.333	0.273	0.360	0.840
1998	0.462	0.223	0.321	0.272	0.335	0.795
1999	0.446	0.192	0.306	0.249	0.315	0.804
Total						
1996	0.164	0.135	0.246	0.190	0.181	0.513
1997	0.129	0.135	0.234	0.185	0.166	0.523
1998	0.111	0.124	0.216	0.170	0.151	0.527
1999	0.105	0.118	0.215	0.167	0.146	0.531

Degree of the Poverty Threat according to the Objective Approach

Dimensions of Poverty Treated Jointly. The household poverty threat in all its dimensions treated jointly has been subject to steady decline in the years under examination. We have observed this decline in all the socio-economic groups with the exception of the ones living on unearned sources other than retirement pay and pension. There was poverty threat growth in this group in 1997. Nevertheless, the membership function values in the poverty sphere declined over the subsequent years of the research in this group of households, too. Attention should be given to the growth in the poverty threat in the group of pensioners in 1999 over 1998.

Households living on unearned sources and farmer households belonged to the group of households with the greatest poverty threat in all years under examination.

Income. The poverty threat in terms of income declined steadily at the national level in subsequent years under research (Table 12.1). We have observed a similar situation in most of the socio-economic groups of households under examination. Farmer and pensioner households are an exception. Their membership function values increased in the last year under examination. Growth in the membership function values also occurred in the group of households living on unearned sources in 1997.

Households living on unearned sources and farmer households are the groups with the highest poverty threat in the income dimension, just as in all poverty dimensions considered together.

Housing conditions. The downward trend in the poverty threat in terms of housing conditions at the national level was inexistent until 1998 (Table 12.1). In 1996-1997 the poverty threat was constant. The reason for this was the considerable growth in the lack of independent living, the variable which is the poverty symptom with the greatest weight in housing dimension. The other poverty symptoms featured a decline in value in this period. We have simultaneously observed changes in various directions to the poverty threat in the housing dimension in 1996-1997 in various socio-economic groups of households. This threat declined only in the employee and farmer groups. The growth in the poverty threat in the subsequent year of research occurred only in the group of households living on unearned sources.

The household poverty threat in terms of housing conditions was much less differentiated in the period under examination depending on the socio-economic group than in terms of the income standing. Farmer households and households living on unearned sources were clearly the most threatened by poverty in all the years under examination.

Household affluence. In the entire period under examination there was a decline from year to year in the poverty threat in terms of household affluence at the national level (Table 12.1). We have observed this trend basically in all the socio-economic groups of households distinguished in the research. The exceptions should include the growth in the level of this threat in 1999 in the group of retired people and pensioners and in 1997 in the group of households living on unearned sources. We have observed the largest poverty threat in its dimension as analyzed in the years under examination in the household groups living on unearned sources, retired people, pensioners and farmers.

Degree of the Poverty Threat according to the Subjective Approach

The degree of the poverty threat in households in the subjective approach has been subject to constant growth in the period under examination in contrast to the objective approach (Table 12.1). This means that the growth in household aspirations in terms of material conditions has been much quicker than the growth in actual material conditions. These same trends are visible in the period under examination in the socio-economic groups of households distinguished in the research with the exception of employee and self-employed households. Here the changes to the membership function values in the poverty sphere took different directions in the individual years. An analysis of the degree of household membership in the poverty sphere shows that in the period under examination the households living on unearned sources and pensioners belonged to the group of households under the greatest poverty threat in the subjective approach.

12.4.2 Poverty Nature

The strong poverty threat in Poland in the objective approach was transitory for most households from 1996 to 1999. Of the 11.16% households strongly at risk in at least one year of the research, only 1.68% were in the sphere of a strong poverty threat for the entire period from 1996 to 1999, representing only 15.05% (Table 12.2). The most numerous group of households was strongly at risk in its income dimension. 26.1% of the households had membership function values higher than 0.5 for at least one year, but at the same time only 2.91% had so for the entire research period (11.15% of the households strongly at risk of poverty in the income dimension for at least one year of research). Considerably fewer household groups were strongly at risk in the dimension of housing conditions and household affluence. Nevertheless, this threat was much more of a chronic nature than in the income dimension. Only 10.48% of the households were strongly at risk of poverty in terms of housing conditions for at least one year while 50.67% of them were under a high degree of poverty threat for the entire research period. In the household affluence dimension these households accounted for 13.34% and 20.84%, respectively. They were 9.24% and 35.28%, respectively of the households in total for both dimensions of non-monetary poverty. The considerably higher chronic poverty threat in the non-monetary dimension than in the income dimension occurs, as mentioned above, from greater variability in the timing of household income than in the variability of housing conditions and household affluence, which are resources.

The high poverty threat according to the subjective approach was more extensive and of a more chronic nature than in the objective approach. According to the households' evaluations 54.26% of the population under examination was in the sphere of strong poverty for at least one year, while 12.89% was in this sphere for all of the years under examination, which accounts for 23.76% of the households under a strong threat for at least one year.

Table 12.2. Poverty Nature in 1996-1999

Poverty dimensions	Percentage of households in high poverty threat*				
	0 years	1 years	2 years	3 years	4 years
Objective approach	88.84	5.02	2.69	1.77	1.68
income	73.90	11.45	7.32	4.42	2.91
housing conditions	89.52	2.34	1.32	1.50	5.31
household affluence	86.66	5.81	2.73	2.02	2.78
non-monetary	90.76	2.76	1.49	1.73	3.26
Subjective approach	45.74	18.08	12.27	11.02	12.89

*Membership function is not less than 0.50 in the objective approach and households which evaluated their material situation as bad or rather bad in the subjective approach.

In the period under examination there were relatively frequent changes in the degree of the poverty threat to households in its various dimensions. The mobility measures estimated on the basis of the transition matrices depict a regular evaluation of these changes (Table 12.3).

More than 23% of the households changed their class of poverty threat in 1997 in comparison with 1996 in the subjective approach. In the consecutive years of the research household mobility by the degree of the poverty threat fell considerably (to 17.3% in 1998 in comparison with the previous year). The percentage of households in which there was a substantial decline in the poverty threat in the objective approach from year to year (shifting to classes with a lower degree of poverty threat) was higher in each one of the years under examination than the percentage of households exhibiting a significant increase in the poverty threat (shifting to classes with a higher degree of poverty threat). These differences, however, declined over subsequent years. The same downward trend in the differences in all the dimensions of objective poverty have also been observed.

Table 12.3. Household Mobility by Membership in Classes of Poverty Threat in 1996-1999

Poverty dimension	Poverty dimension					
	objective approach					subjective approach
	income	housing conditions	household affluence	non-monetary	total	
1996-1997:						
SU	22.61	13.11	26.65	16.87	23.56	40.98
SU^+	14.20	4.86	15.27	9.06	14.68	20.50
SU^-	8.41	8.25	11.38	7.81	8.88	20.48
SU^+-SU^-	5.80	-3.39	3.89	1.25	5.80	0.02
1997-1998:						
SU	18.47	7.77	23.00	13.63	18.62	35.70
SU^+	10.62	5.25	15.62	9.82	11.52	16.40
SU^-	7.85	2.52	7.38	3.81	7.10	19.30
SU^+-SU^-	2.77	2.73	8.24	6.01	4.42	-2.89
1998-1999:						
SU	16.62	6.10	21.19	12.23	17.30	36.91
SU^+	9.08	4.12	11.09	6.79	9.59	16.46
SU^-	7.54	1.98	10.10	5.44	7.71	20.45
SU^+-SU^-	1.54	2.14	0.98	1.35	1.88	-3.98

The most frequent changes to the degree of the poverty threat in the objective approach were in the household affluence dimension (they affected from 26.65% of the households in 1996-1997 to 21.19% of the households in 1998-1999). They were positive in all the years. In 1998-1999 the percentage edge enjoyed by households in which the poverty threat degree fell in households featuring an increase in the poverty threat amounted to just 0.98%. In the income dimension we have also observed an edge enjoyed by positive changes in the poverty threat over negative changes in each of the years under examination. These differences amounted to 5.80% in 1996-1997 and to 1.4% in 1998-1999. The smallest changes in the degree of the poverty threat in the objective approach took place in the years under examination in the housing conditions dimension. Only 13.11% of the households in 1996-1997 and 6.10% of the households in 1998-1999 changed their class of poverty threat in this period. In the period of 1996-

1997 the number of households in which there was considerable growth in the poverty threat surpassed the number of households with a substantial decrease in the poverty threat. This was the only case of negative change to the poverty threat in the objective approach. This was caused by a significant decrease in the independence of household residence in these years.

Household mobility by the poverty threat in the subjective approach was much greater than in the objective approach. 40.98% of the households changed the class of the poverty threat risk in 1996-1997 (during the period of the greatest changes) in the subjective approach while the changes in the objective approach in this period affected only 23.30% of households. The feelings of households with regard to their poverty threat were therefore subject to much more fluctuation than the actual changes in their material conditions indicate. The difference between positive and negative changes increased in the subjective approach from year to year in favor of negative changes. This shows the growing pessimism among the households under research with respect to their material standing.

12.4.3 Poverty Determinants

The parameters of the probit model were estimated using the maximum likelihood method. The comparison of the empirical values of the χ^2 statistic with the corresponding critical values points to a high goodness of fit of both models and the significance of all the independent variables appearing in them (the distinguished variants of the attributes) treated jointly. When researching the significance of the individual independent variables in the models, a significance level equal to 0.05 was accepted. (Tables 12.4 and 12.5).

Socio-economic Group. The benchmark for evaluating the impact exerted by a household belonging to a given socio-economic group risking poverty was the group of employee households. This means that the degree of the poverty risk to household groups distinguished by socio-economic group was considered against the benchmark of the level of the risk to the group of employee households. In both the objective and subjective approaches the household groups with the highest risk of poverty are households living on unearned sources other than retirement pay and pension. Nevertheless, in the subjective approach, the parameter in the second group of households mentioned is significant only at a level of significance equal to 0.64. By far, the lowest risk of a strong poverty threat is present in both measurement approaches in the group of self-employed households.

Table 12.4. Probit Estimates of High Poverty Threat Risk in 1999. Objective Approach

Group of variables and variable	Estimate	Standard error	t-statistic	p-value
Socio-economic group:				
Employees	0	-	-	-
Employee-farmers	0.075	0.097	0.776	0.438
Farmers	0.287	0.107	2.681	0.007
Self-employed persons	-0.602	0.218	-2.765	0.006
Retired people	0.193	0.095	2.028	0.043
Pensioners	0.534	0.102	5.257	0.000
Persons living on unearned sources	0.835	0.131	6.368	0.000
Number of persons in a household:				
1	0.346	0.090	3.856	0.000
2	0	-	-	-
3	-0.141	0.091	-1.547	0.122
4	-0.209	0.097	-2.160	0.031
5	-0.028	0.102	-0.275	0.783
6 and more	0.216	0.106	2.046	0.041
Place of residence:				
Cities: more than 500 thousand inhabitants	0	-	-	-
Cities: 200-500 thousand inhabitants	0.306	0.168	1.815	0.070
Cities: 100-200 thousand inhabitants	0.283	0.194	1.464	0.143
Cities: 20-100 thousand inhabitants	0.330	0.162	2.033	0.042
Cities: less than 20 thousand inhabitants	0.531	0.168	3.168	0.002
Rural area	1.118	0.150	7.462	0.000
Household head's education:				
Elementary or lower	1.560	0.232	6.714	0.000
Basic vocational	1.060	0.233	4.550	0.000
Secondary	0.592	0.239	2.477	0.013
University	0	-	-	-
Household head's age:				
Less than 25 years	0.240	0.151	1.595	0.111
25-34 years	0.322	0.085	3.810	0.000
35 years and more	0	-	-	-
Household status on labour market:				
At least 1 unemployed person	0.615	0.085	7.262	0.000
No unemployed persons	0	-	-	-
Disabled persons:				
At least 1 disabled person	-0.223	0.063	-3.524	0.000
Without disabled persons	0	-	-	-

Chi-Squared (28) 877,230
Significance Level 0.000

Table 12.5. Probit Estimates of High Poverty Threat Risk in 1999. Subjective Approach.

Group of variables and variable	Estimate	Standard error	*t*-statistic	*p*-value
Socio-economic group:				
Employees	0	-	-	-
Employee-farmers	-0.203	0.105	-1.926	0.054
Farmers	-0.691	0.141	-4.906	0.000
Self-employed persons	-1.063	0.282	-3.763	0.000
Retired people	-0.421	0.093	-4.515	0.000
Pensioners	0.177	0.095	1.853	0.064
Persons living on unearned sources	0.482	0.127	3.812	0.000
Number of persons in a household:				
1	0.222	0.091	2.461	0.014
2	0	-	-	-
3	0.056	0.092	0.606	0.544
4	0.127	0.097	1.302	0.193
5	0.311	0.108	2.885	0.004
6 and more	0.516	0.117	4.407	0.000
Place of residence:				
Cities: more than 500 thousand inhabitants	0	-	-	-
Cities: 200-500 thousand inhabitants	-0.124	0.111	-1.116	0.265
Cities: 100-200 thousand inhabitants	-0.053	0.133	-0.397	0.691
Cities: 20-100 thousand inhabitants	-0.243	0.109	-2.235	0.025
Cities: less than 20 thousand inhabitants	-0.241	0.119	-2.026	0.043
High Rural area	-0.596	0.108	-5.508	0.000
Household head's education:				
Elementary or lower	0.709	0.157	4.522	0.000
Basic vocational	0.611	0.154	3.980	0.000
Secondary	0.301	0.156	1.931	0.054
University	0	-	-	-
Household head's age:				
Less than 25 years	0.132	0.144	0.920	0.358
25-34 years	-0.052	0.086	-0.603	0.547
35 years and more	0	-	-	-
Household status on labour market:				
At least 1 unemployed person	0.277	0.085	3.267	0.001
No unemployed persons	0	-	-	-
Disabled persons:				
At least 1 disabled person	0.053	0.064	0.830	0.406
Without disabled persons	0	-	-	-
Income:				
High value of m. f. [a]	0	-	-	-
Low value of m. f.	-0.701	0.070	-10.025	0.000

Table 12.6. (cont)

Housing conditions:				
High value of m. f. [b]	0.320	0.091	3.523	0.000
Average value of m. f.	0	-	-	-
Low value of m. f. [b]	-0.030	0.081	-0.371	0.711
Household affluence:				
High value of m. f. [b]	0.689	0.066	10.431	0.000
Average value of m. f.	0	-	-	-
Low value of m. f. [b]	-1.147	0.463	-2.476	0.013

Chi-Squared (23) 916,871
Significance Level 0.000
[a] Membership function >0.5;
[b] 10% of the households under the greatest and least risk of poverty were included among the groups of households with high and low membership function values

Number of Persons in a Household. The benchmark for evaluating the impact of the number of household members risking a household being in poverty was a two-person household. The number of persons significantly determined the presence of a high poverty threat in the objective and subjective approaches in single and multiple-person households. Moreover, the membership in the group of four-person households has an important, albeit negative impact in the objective approach on the risk of a strong poverty threat. Thus the determinants of poverty in both measurement approaches are similar. At the same time, the growth in the number of persons in multiple-person households increases the risk significantly. The causes of this situation should be seen in the fact that multiple-person households are most frequently established by families with multiple children in which most persons are not working professionally. In turn, in single-person households the fixed costs of maintenance are considerably higher than in households with a greater number of persons, which as a result increases the risk of a strong poverty threat.

Class of Place of Residence. To evaluate the impact exerted by the place of residence on the poverty risk, the accepted benchmark was a household residing in the largest city (above 500 thousand inhabitants). In both the objective and subjective approaches the highest risk of poverty is significantly increased by residence in rural areas and small cities (below 100 thousand inhabitants).

At first glance the different hierarchy of the impact exerted on the risk of high poverty in both measurement approaches may appear to be surprising. In the objective approach the risk of a high poverty threat increases as the size of the household's place of residence declines (it is decisively the highest in the rural areas). In the subjective approach in turn this risk in the rural areas is decisively lower than in cities. Households evaluate their poverty threat relatively by comparing their material standing to other households in their place of residence. Rural households feature much less variation in the living standard than households residing in cities; that is why their evaluation of the poverty threat are relatively low.

Household Head's Education. The level of the household head's education significantly determines the risk of being in poverty in both the objective and subjective approach. The benchmark for evaluating the impact exerted by the levels of the household head's education distinguished in the model on the poverty risk was the group of households whose head has a university level education. The risk of a high poverty threat clearly decreases in both measurement approaches as the level of the household head's level of education increases.

Household Head's Age. The group of households whose head is 35 years and older has been selected as the benchmark for these analyses. The differences in the level of risk of a high poverty threat between this group of households and others distinguished by the household head's age are not basically significant. The group of households with heads aged 25-34 forms an exception in the objective approach with a high risk of a strong poverty threat.

Household's Status on the Labor Market. A group of households with no unemployed persons and a group with at least one unemployed person were distinguished by their status on the labor market. The estimates obtained of these parameters point to considerable risk of a high poverty threat among households with unemployed persons both in the objective and in the subjective approach.

Disabled Persons in the Household. Households have been divided into two groups of households: with and without disabled persons. The presence of disabled persons in households significantly increases the risk of a high poverty threat in both measurement approaches.

High Poverty Threat in Its Objective Dimensions. Membership function values have also been incorporated in the model as potential determinants of poverty in the subjective approach in the sphere of poverty in its objective dimensions. High membership function value in all objective measures of poverty significantly increase the risk of a high poverty threat in the subjective approach. The strongest impact is present in the household affluence dimension; the weakest impact is present in housing conditions.

The possession of durable goods is the most visible, external indicator of material standing in the evaluation of households.

12.5 Summary

The degree of the poverty threat in Poland from 1996 to 1999 in the objective approach showed a downward trend in all its dimensions. Nevertheless in 1999 disquieting signals of possible growth in the poverty threat in forthcoming years began to appear, viz. growth in the poverty threat in distinguished dimensions in some socio-economic groups of households. The greatest poverty threat was clearly present in this measurement in the area of household affluence.

The poverty threat in the period under analysis according to the feelings of households was not only higher than indicated by the objective approach measurement but it also grew steadily. This points to a considerably quicker growth in the aspirations of households with respect to their material conditions than the growth in their actual material conditions.

The strong poverty threat in Poland according to the objective approach was transitory for most households in the period under research. Nevertheless, this situation may undergo unfavorable changes in upcoming years. The decline in economic growth and the unemployment growth that accompanies it, which, as research has shown, is one of the basic determinants of the high threat of poverty, may contribute to a reversal in the favorable trends of 1996-1999. The high poverty threat according to the subjective approach in the period under examination had a much broader extent and a more chronic nature than in the objective approach. Household mobility by the poverty threat degree in the subjective approach was also much greater than in the objective approach. Moreover, negative changes prevailed over positive changes in this measurement approach, in contrast to what occurred in the objective approach.

The household head's level of education, the household's status on the labor market and the presence of disabled persons in the household were the most influential factors affecting the risk of a high poverty threat in both measurement approaches. Moreover, in the subjective approach, the risk of high poverty was materially exacerbated by high membership function values in the poverty sphere in the objective approach dimensions distinguished in the research.

The findings of the analysis confirm the hypothesis that poverty in Poland has many non-overlapping dimensions and that its measurement exclusively from an income standing vantage point is highly insufficient.

This points to the need for further poverty threat research applying the multidimensional approach and using panel data as a basis. The findings of such research should form a valuable source of information for articulating the objectives and methods of executing social policy to counteract poverty.

References

Betti G (1996) A Longitudinal Approach to Poverty Analysis: the Latent Class of Markov Model. Statistica 56:345-359

Błaszczak-Przybycińska I, Kotowska IE, Panek T, Podgórski J, Rytelewska G, Szulc A (1998) Living Conditions of Polish Households in 1995-1995. Current State, Threats, Opportunities. Institute of Pubic Affairs, Warsaw

Błaszczak-Przybycińska I (1992) Multidimensional Statistical analysis of Poverty. In: Poverty Measurement for Economies in Transition in Eastern European Countries. Polish Statistical Association, Warsaw, pp 307-327

Cerioli A, Zani S (1990) A Fuzzy Approach to the Measurement of Poverty, In: Dagum C, Zenga M (eds) Income and Wealth Distribution, Inequality and Poverty. Studies in Contemporary Economics. Springer Verlag, Berlin, pp 272-284

Cheli B, Ghellini G, Lemmi A, Pannuzi N (1994) Measuring Poverty in the Countries in Transition via TFR Method: The case of Poland in 1990-1991. Statistics in Transition 3:585-636

Deniszczuk L, Sajkiewicz B (1996a) Category of Social Minimum. In: Golinowska S (ed) Polands' Poverty. Criteria. Valuation. Counteracting (in Polish). Institute of Labor and Social Affairs, Warsaw, pp 146-185

Deniszczuk L, Sajkiewicz B (1996b) Category of Subsistence Minimum. In: Golinowska S (ed) Polands' Poverty. Criteria. Valuation. Counteracting (in Polish). Institute of Labor and Social Affairs, Warsaw, pp 18-40

Dubois D, Prade H (1980) Fuzzy Sets and Systems: Theory and Applications. Academic Press, New York

Polish Central Statistical Office (1999) Methodology of the Household Budget Survey (in Polish). Warsaw

Greene WH (1997) Econometric Analysis. Prentice Hall, New York

Panek T (1994) A Multidimensional Analysis of the Sphere of Poverty in Poland in 1990-1992. RECESS Research Bulletin 2:43-46

Panek T (1996) A System to Monitor the Living Conditions on Polish Households. Statistics in Transition 2:979-1004

Panek T, Podgórski J, Szulc A (1999) Poverty: Methodology and Practice of Measurement (in Polish). Warsaw Scholl of Economics, Warsaw

Rodgers JR, Rodgers JL (1993) Chronic Poverty in the United States. The Journal of Human Resources 28:25-54

Shorrocks AF (1978) The Measurement of Mobility. Econometrica 46:1013-1024

Stevens AH (1999) Climbing Out of Poverty, Falling Back In. The Journal of Human Resources 34:557-588

Warsaw School of Economics (2000) Polish International Economic Report 2000/2001, Warsaw

Zadeh LA (1965) Fuzzy Sets. Information and Control 8:338-353

13 Modelling Fuzzy and Multidimensional Poverty Measures in the United Kingdom with Variance Components Panel Regression

Gianni Betti, Antonella D'Agostino, Laura Neri

Gianni Betti, Laura Neri
Dipartimento Metodi Quantitativi, University of Siena
Antonella D'Agostino
Dipartimento di Statistica e Matematica per la Ricerca Economica, University of Naples "Parthenope"

13.1 Introduction

In this Chapter we propose a methodology to model fuzzy and multidimensional poverty measures in order to study poverty dynamics and influencing socio-demographic factors.

A large amount of literature exists which refers to i) the study of fuzzy and multidimensional poverty in a cross-sectional context, and ii) the study of non-fuzzy poverty dynamics. Not many studies have been carried out on iii) fuzzy and multidimensional measures in a longitudinal perspective.

Concerning category i) above, over the last decades many studies have paid increasing attention to the multidimensional aspects of the phenomenon of poverty and living conditions. These aspects are not taken into account in the so called traditional approach to poverty analysis which only considers monetary indicators (e.g. income or consumption expenditure); in this context the theory of fuzzy sets has been introduced by Cerioli and Zani (1990) and developed by Cheli and Lemmi (1995) in order to overcome some limitations of the traditional approach and in order to define multidimensional fuzzy poverty measures.

A large amount of literature also exists which refers to the study of non-fuzzy poverty dynamics; one of the first contributions, by Lillard and Willis (1978) concentrated on earning dynamics using variance-component models, applied to the Panel Study of Income Dynamics (PSID). More recently, Stevens (1999) has compared duration models with variance component models using an updated set of the PSID. Jenkins (2000) describes a wide range of multivariate models of income and poverty dynamics, including: *i)* longitudinal poverty pattern models, *ii)* transition probability models, *iii)* variance component models, *iv)* structural models, with an application to the first 6 waves of the British Household Panel Survey (BHPS). Devicenti (2001) starting from Stevens methodology (1999) studies poverty dynamics in Great Britain from 1991 to 1997.

The first attempt to study poverty dynamics in a longitudinal context was due to Cheli (1995) extending the Totally Fuzzy and Relative (TFR) approach to two periods; this was later developed by Cheli and Betti (1999) and Betti, Cheli and Cambini (2004) introducing fuzzy transition matrices and dynamic indices. Betti and Verma (1999, 2004) have focused more on capturing the multi-dimensional aspects, developing the concepts of "manifest" and "latent" deprivation to reflect the intersection and union of different dimensions; this approach has been further developed and extended to the longitudinal perspective in a more general and consistent way, in Chapter 6 of the present Book. The first attempt to model fuzzy poverty dynamics by means of panel regression is the application of Betti, D'Agostino and Neri (2002) to the first 7 waves of the BHPS. This Chapter is an extension of the latter work: here our original contribution can be summarized as follows:

a) Adoption of a new definition for the membership function to the subset of poor (see Sect. 13.2), in line with the developments reported in the first part of the Book.

b) Extension of the analysis from 7 to 12 waves. Moreover since the BHPS questionnaire has been enriched from wave 6 onwards, two distinguished analyses have been performed: the first based on a small set of common qualitative (supplementary) indicators available for the whole period and the second based on a larger set of indicators present in waves from 6 to 12 only.

c) Particular attention is also paid to the model specification of the time effect; starting from a very flexible parameterisation by means of time indicator variables we look for the best specification of the time effect and then introduce socio-demographic covariates into the model.

The Chapter is organised as follows. In Sect. 13.2 two different measures for the definition of the concept of poverty are presented. The panel regression models are presented in Sect. 13.3. The empirical analysis is reported in Sect. 13.4 (cross-sectional) and in Sect. 13.5 (poverty dynamics); it is based on the data set collected by the BHPS from 1991 to 2002. Finally some concluding remarks are made in Sect. 13.6.

13.2 Fuzzy and multidimensional poverty definitions

As shown in the first part of the present Book, the adoption of a multidimensional approach leads to two main problems: the choice of the indicators and the aggregation process. Although deprivation is widely recognized as a multidimensional phenomenon, we still believe that indicators based on monetary variables have a fundamental role and therefore are worthy of special treatment. For this reason two different fuzzy measures are considered: the first one is based only on a monetary variable and here it is referred to as Fuzzy Monetary (FM); the second measure is based on several indicators relating to housing conditions, durable goods, etc... and here it is referred to as Fuzzy Supplementary (FS).

The monetary variable used for the FM method consists in the net equivalised household income y. Making use of the concepts of the fuzzy set theory, the degree of deprivation of any household *i* at any period *t* is defined as the membership function to the fuzzy set of poor:

$$\mu_{it}=FM_{it}=\left(\sum_{\gamma=i+1}^{n} w_{\gamma t}y_{\gamma t}\Big/\sum_{\gamma=2}^{n} w_{\gamma t}y_{\gamma t}\right)^{\alpha}\;;\quad \mu_{nt}=0 \tag{13.1}$$

As proposed by Cheli and Lemmi (1995) we determine parameters α_t so that the membership function means are not merely equal to 0.5, but are equal to the proportion of poor units according to the traditional approach (the so called head count ratio H).

The FS measure is based on some supplementary variables x_{ijt} (j = 1, 2, . . ., k), such as amenities in the household, ability to afford durable goods, accommodation problems, and any other variables relevant for the multidimensional definition of deprivation. The construction process of this measure is fully described in Betti and Verma (1999). When supplementary variables are ordinal with two or more categories, for each variable j, with ordered categories from 1 (least deprived) to M (most deprived), the single poverty indicator for all households in category m is defined as follows:

$$s_{ijt}=\frac{m-1}{M-1} \tag{13.2}$$

When supplementary variables are quantitative poverty indicators they can be calculated in a way similar to formula (13.1). The aggregation process of the single indicators into the multidimensional measure is constructed as a weighted mean:

$$s_{it}=FS_{it}=\frac{\sum_{j=1} w_j s_{ijt}}{\sum_{j=1} w_j} \tag{13.3}$$

The weights w_j are determined by two statistical considerations: *i)* firstly, the weight is determined by the power of the variable to "discriminate" among individuals in the population, that is, by its dispersion, measured by its coefficient of variation; *ii)* in order to avoid redundancy, it is necessary to limit the influence of those characteristics that are highly correlated to the others. For a detailed description of the weighting procedure see Betti and Verma (1999).

13.3 Panel regression models with variance components

In most applications, data sets present multiple measurements over time on the same statistical unit. Regression model is assuming independent errors are not appropriate in this case because repeated measures can be correlated. Hence modelling an appropriate covariance structure is essential so that inferences about means are valid. Therefore, the parameters of the mean model are referred to as *fixed-effect parameters* (associated with known explanatory variables) and as *random-effect parameters* (associated to the chosen covariance structure).

In the present framework, dependent variables are $\mathrm{FM}_{\mathrm{it}}$ and $\mathrm{FS}_{\mathrm{it}}$ respectively the poverty indicator based on income (FM measure) and the poverty indicator based on qualitative variables (FS measure) at time t (t = 1,…,T) for each statistical unit i (i = 1,…n). The following logit transformation:

$$\mathrm{y}_{\mathrm{it}}^{(\mathrm{FM})}=\mathrm{logit}\left(\mathrm{FM}_{\mathrm{it}}\right);\quad \mathrm{y}_{\mathrm{it}}^{(\mathrm{FS})}=\mathrm{logit}\left(\mathrm{FS}_{\mathrm{it}}\right) \tag{13.4}$$

is performed in order to create two variables ranging between $-\infty$ and $+\infty$. The statistical model for both indicators is specified as:

$$y^{(\cdot)}{}_{it} = \boldsymbol{\beta}'\mathbf{x}_{it} + \phi_t + u_{it}\,. \tag{13.5}$$

Here $\boldsymbol{\beta}$ is an unknown vector of fixed-effect parameters associated to k time-varying exogenous variables $\mathbf{x}_{it}$ on individual i, ϕ_t is a parametric or non parametric specification for the time effect and u_{it} is the error structure that takes into account correlation among repeated measures. We assume that:

$$u_{it} = \delta_i + \xi_{it}\,, \tag{13.6}$$

where $\delta_i \approx N\left(0,\sigma_\eta^2\right)$ and ξ_{it} follows a AR(1) structure, e.g.:

$$\xi_{it} = \rho\xi_{it-1} + \eta_{it}\,. \tag{13.7}$$

Here η_{it} is a purely random component assumed to be *i.i.d.* as $N\left(0,\sigma_{\eta}^{2}\right)$ and ρ is the serial correlation coefficient common to all statistical units. We also assume that δ_i and η_{it} are independent of each other and of $\mathbf{x}_{it}$ and ϕ_t (Lillard and Willis, 1978).

Note that this is a particular specification of covariance error structure that combines autoregression with variance component so as to obtain a model allowing for both heterogeneity and autocorrelation (Anderson and Hsiao, 1981; Mansour et al. 1985; Goldstein et al. 1994).

The model assumes that there are two random effects: one (δ_i) is assumed to persist through the period of observation (called, for this reason, permanent variation); and the other one (ξ_{it}) has the desired property of correlations, being larger for nearby times than far-apart times. In order to explain the amount of variation due to permanent component δ_i, the intra-class correlation coefficient can be computed as:

$$\hat{\gamma} = \hat{\sigma}_{\delta}^{2} \big/ \hat{\sigma}_{u}^{2} \,. \tag{13.8}$$

However, in this specification the assumption concerning the initial observation plays a crucial role in interpreting the model and in devising consistent estimates. Therefore for the first response on each unit it is assumed that $y_{i0} \sim N\left(\boldsymbol{\beta}'\mathbf{x}_{i0}, \frac{\sigma_{\eta}^{2}}{1-\rho^{2}}\right)$. For t >1 the residual covariance structure is (Anderson and Hsiao 1982):

$$E\left(u_{it}u_{\gamma t'}\right) = \begin{cases} \sigma_{\delta}^{2} + \dfrac{\sigma_{\eta}^{2}}{1-\rho^{2}} & \mathrm{i} = \gamma \quad \mathrm{t} = \mathrm{t}' \\ \sigma_{\delta}^{2} + \rho^{s}\dfrac{\sigma_{\eta}^{2}}{1-\rho^{2}} & \mathrm{i} = \gamma \,, \left|\mathrm{t}\text{-}\mathrm{t}'\right| = \mathrm{S} > 0 \\ 0 & \mathrm{i} \neq \gamma \end{cases} \,. \tag{13.9}$$

13.4 Cross-sectional empirical analysis

The empirical analysis has been conducted using twelve waves of data of the British Household Panel Survey (BHPS) for the years 1991 – 2002[1].

The BHPS was designed as an annual survey of each adult (16+) member of a nationally representative sample of more than 5,000 households, making a total of approximately 10,000 individual interviews annually. The same individuals are re-interviewed in successive waves and, if they split-off from original households, all adult members of their new households are also interviewed. Children are interviewed once they reach the age of 16. Thus the sample is expected to remain broadly representative of the population of Britain as it changes through the reference period, except for the effect of immigration and panel attrition.

The derived BHPS data set we work with is the one used by Bardasi et al. (2004); this data set reports incomes defeated to January 2003 prices.

The sample used to construct the household poverty indicators (see formulas (13.1) and (13.3)) consists of those households in which all eligible adults gave a full interview in each wave. In this data set the net equivalised household income[2] is present for all individuals; missing values in the supplementary variables have been imputed using the approach adopted by Raghunathan et al. (2001). For the reference year 1991, the poverty line has been calculated as half of the mean net equivalent household income; the line results as being equal to £ 153.17 per week among the 4826 households. Table 13.1 reports the percentages of poor households in waves 1-12 according to the traditional approach (the head count ratios H_t) and the values of parameters α_t of formula (13.1) so that:

$$E\left[FM_{it}\right]=H_t \tag{13.10}$$

Therefore the head count ratios coincide with the household membership function means calculated year-by-year. These show a declining behaviour pattern from 1991 to 1998, and from 1999 to 2002, while there is a slight increase between 1998 and 1999.

Note that in order to identify the year-by-year household head count ratios H_t, the poverty line in the analysis presented here has been calculated for the first period only and is kept fixed (in real terms) for the following years.

[1] Wave number 13 is currently available but it was not so during the data analysis.

[2] This is the sum of all individual net incomes deflated by the McClements (1977) equivalence scale.

Table 13.1. Cross-sectional membership functions

Wave	1991	1992	1993	1994	1995		
$E[FM_{it}]=H_t$	0.2075	0.1690	0.1659	0.1588	0.1460		
α_t	1.9515	2.2065	2.2272	2.2580	2.3825		
$E\left[FS_{it}^{1\text{-}12}\right]$	0.2536	0.2415	0.2286	0.2136	0.2039		
N	4826	4556	4354	4378	4259		
Wave	1996	1997	1998	1999	2000	2001	2002
$E[FM_{it}]=H_t$	0.1323	0.1284	0.1190	0.1230	0.1113	0.0974	0.0921
α_t	2.4981	2.6155	2.5357	2.6182	2.7943	2.8602	2.8578
$E\left[FS_{it}^{1\text{-}12}\right]$	0.1864	0.1751	0.1636	0.15895	0.1424	0.1312	0.1215
$E\left[FS_{it}^{6\text{-}12}\right]$	0.0793	0.0712	0.0657	0.0660	0.0623	0.0575	0.0555
N	4372	4384	4328	4273	4194	4104	3969

In order to evaluate the household membership functions according to the FS measure (formula (13.3)) two sets of supplementary variables are considered. The first set ($FS_{1\text{-}12}$) consists of those variables collected in the entire time period (waves 1 to 12) and are denoted with an asterisk in Table 13.2: they refer to housing conditions and to the presence of durable goods. The exhaustive list of poverty symptoms is: house which is not owned; and lack of central heating, colour TV, video recorder, washing machine, dishwasher, home computer, CD player, microwave, car or van. The second set ($FS_{6\text{-}12}$) consists of a larger group of variables collected from wave 6 onward; the complete list is reported in Table 13.2.

It should be noted that the indicators reported in Table 13.2 are not proper poverty symptoms: sometimes, it could merely be a matter of choice whether to own a car or not (especially if someone lives in Central London); therefore it would be more informative to know whether or not someone can afford a particular good *if they wanted it*. Unfortunately, this information is not collected by the BHPS, at least in the first waves.

Let us now analyse household means of the two FS indicators reported in Table 13.1: in this case we can observe a regular decrease of the indicator $FS_{1\text{-}12}$ over twelve years, while the indicator $FS_{6\text{-}12}$ shows a slight increase between 1998 and 1999.

Table 13.2. Supplementary indicators

Variable name	Variable label	Variable name	Variable label
wHSWND	House owned	wHSPRBO	Accommodation has rot in windows, floor
wHSCANA	Keep home adequately warm	wHSPRBP	Pollution and environmental problems
wHSCANB	Pay for annual holiday	wHSPRBQ	Vandalism or crime
wHSCANC	Replace furniture	wHSGDN	Accommodation has a terrace or garden
wHSCANE	Buy new clothes	wFNCARS*	Car or van for private use
wHSCAND	Eat meat on alternate days	wHEATCH*	House has central heating
wHSCANF	Feed visitors once a month	wHCD1USE*	Colour TV
wHSKCH	Accommodation has a separate kitchen	wHCD2USE*	VCR Video recorder
wHSBTH	Accommodation has a separate bathroom	wHCD3USE*	Washing machine
wHSTLT	Accommodation has an indoor toilet	wHCD6USE*	Dish washer
wHSPRBG	Accommodation has shortage of space	wHCD7USE*	Microwave oven
wHSPRBH	Accommodation has noise from neighbours	wHCD8USE*	Home computer
wHSPRBI	Accommodation has street noise	wHCD9USE*	CD player
wHSPRBJ	Accommodation has not enough light	wHCD10USE	Satellite dish
wHSPRBK	Accommodation has lack of adequate heating	wHCD11USE	Cable TV
wHSPRBL	Accommodation has condensation problem	wHCD12USE	Telephone in accommodation
wHSPRBM	Accommodation has leaky roof	wXPHSD1	Housing payment required borrowing
wHSPRBN	Accommodation has damp walls, floors		

- The analysis conducted on waves 1-12 is based on those variables denoted by the asterisk.

13.5 Longitudinal empirical analysis

BHPS identifies the individual person, as the "unit of analysis" and establishes such a rule independently of the phenomenon being studied. However, this Chapter deals with the multidimensional aspects of poverty dynamics, and in this context there is no unanimity in the choice of the longitudinal unit; the controversy is about choosing individuals or households. Considering the unit identification problem, it seems to be reasonable to consider the individual person as a unit of the analysis: persons remain identifiable over periods, but the identification of families is complicated by marriages, divorces, births and deaths, and movement of individual family members (Trivellato, 1998).

On the other hand, it is reasonable to associate poverty dynamics to household variables rather than to individual characteristics. Particularly in the FS approach which is based on supplementary variables such as housing conditions, the presence of some durable goods, etc.; moreover in the FM approach the poverty indicator is computed considering the net equivalised household income which also depends on the size and composition of the whole household. Furthermore, the choice of the individual as the longitudinal unit with household variables generates some complicated econometric problems concerning the specification of the models, already introduced in Sect. 13.3, specifically the effect of: *i)* presence of correlation among members sharing the same household over time; *ii)* introduction of different individual effects for units having exactly the same values for the dependent variable and covariates.

For these reasons, the present analysis is based on the *longitudinal household.* In order to follow a complex unit such as the household, the definition of a set of follow-up or tracing rules becomes more and more important. These rules can be simple for individuals sharing the same household across the reference period; in the other cases, it becomes more complicated to construct rules because of longitudinal changes. There are different ways for defining the longitudinal household. In the present analysis we have chosen to follow adult individuals in the current household for each wave, even if they split-off from the original household; for a detailed discussion about the longitudinal unit of analysis see Betti, D'Agostino and Neri (2002).

The analysis refers to the unbalanced panel of longitudinal households considering different reference time periods separately: i) the whole panel, waves 1-12, in this case the definition of the FS indicator concerns only the subset of indicators common to the twelve waves; here the total sample size consists of 51936 repeated measurements. ii) For waves 6-12, the FS indicator concerns a richer set of indicators common to these final waves; here the total sample size consists of 29624 repeated measurements. For details on the variables involved in the two analyses see Table 13.2.

The models specified in (13.4) have been estimated and in each model the dependent variable consists, alternatively, of one of the two poverty indicators. In order to compare results of the parameter estimation they have been standardized.

The time indicator is specified in two different ways: a linear specification, with the reference variable TREND; and alternatively with a non parametric specification based on dummy variables (DT1 – DT12).

13.5.1 Trend estimation

Particular attention is given to the specification of time effect. For this reason we estimate the models specified in formula (13.4) without covariates. The time effect is introduced in a non parametric form in order to obtain a more flexible model, using dummy variables as time indicators in each wave. Four different models have been estimated taking into account the two data sets described in Sect. 13.5 and the two poverty indicators FM and FS. The estimated trends are plotted in Figures 13.1 and 13.2, and as expected, both measures show a decreasing level of poverty, whichever data set is used[3].

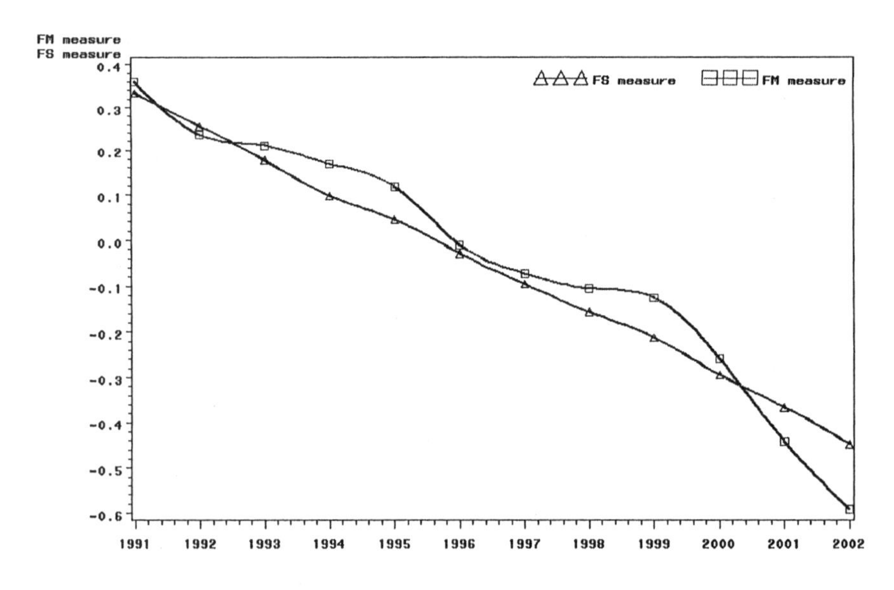

Fig. 13.1. Estimated trends in waves 1 to 12

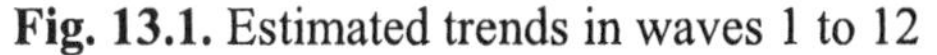

[3] Note that this decline does not necessary imply a decreasing level of relative poverty, defined in relation to poverty lines determined independently at each wave. In this analysis, the poverty line is anchored at year 1991, though income amounts have been adjusted for price inflation. Also note that since the FM measure is defined in the same way for both the periods, obviously the shape drawn, using dataset waves 6-12 and the dataset waves 1-12 is the same between 1996 and 2002.

This generally means a higher improvement of living conditions over time. Moreover the shapes are very different and depend on the two measures and on the two data sets used. Results are very interesting as follows.

Using waves 1-12, the FS measure substantially shows a linear trend; on the contrary, the FM measure needs time indicator dummies to properly capture its modulation over time. In fact, it is quite plausible that income fluctuations make a smoother slope over time than the poverty measure based on supplementary variables that takes into account housing conditions and possession of durable goods. A test based on Akaike's Information Criterion (AIC) suggests reducing parameters estimate for the FS measure by introducing a time dependence specified as a first degree polynomial (AIC_{FM}=107747, AIC_{FS}=87986 with time indicator dummies and AIC_{FM}=108072, AIC_{FS}=87981 with linear trend specification).

Using waves 6-12, both poverty measures show a non-linear trend; the AIC criterion confirms this hypothesis (AIC_{FM}=62917, AIC_{FS}=65514 with time indicator dummies and AIC_{FM}=63079, AIC_{FS}=65576 with linear trend specification).

Differences in the shape of the trend suggest that the choice of supplementary variables can affect the estimated time dependence. In fact, variables included in waves 6-12 represent aspects of life-style and of environmental problems, which are expected to be more variable over time.

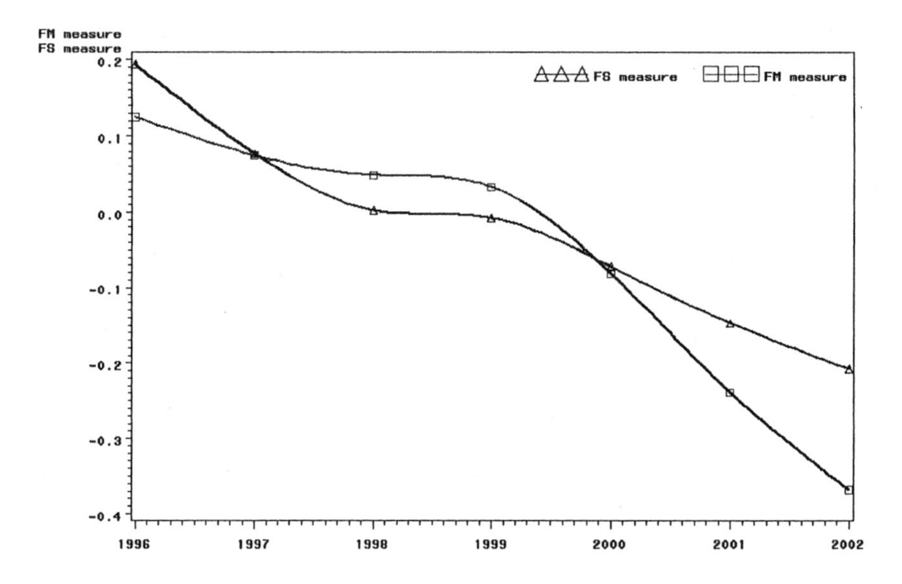

Fig. 13.2. Estimated trends in waves 6 to 12

Let us now consider the random-effect parameters and the autoregressive component. All estimates reported in Table 13.3 are significantly different from zero at least at 1% level. Results for waves 1-12 show the $\hat{\rho}$ estimate for the FS measure to be larger than the one for the FM measure; the opposite result is observed for waves 6-12. The former evidence can be explained by considering that housing conditions and possession of durable goods (which are the main items in the FS measure for waves 1-12) are much less volatile than the monetary variable. On the other hand, the latter evidence leads to reflection on the nature of supplementary variables used for computing the FS measure for waves 6-12: many of these variables (such as environmental and life style indicators) can be much more volatile than income over time, and, on the basis of statistical criteria noted in Sect. 13.2, they receive larger weights than housing conditions and possession of durable goods.

Let us consider the random-effects. We first refer to waves 1-12. The total within-year variance is 0.903 for the FM measure; the amount of variation due to the permanent component (δ_i) is about 43% of total variation (see the column referring to $\hat{\gamma}$ in Table 13.3). The remaining 57% is due to purely stochastic variation, from period to period, and to serial correlation contained in the component ξ_{it}. In the case of the FS measure, the total within-year variance is 0.846, and about 59% of it is explained by permanent variation. This last value is higher than the one computed for the FM measure and it confirms the evidence already highlighted in the case of the trend model. The effect of the purely stochastic variation ($\hat{\sigma}_\eta^2$) for the FM measure is stronger than the one for the FS measure; it seems to be reasonable given that monetary variables can be more affected by measurement errors.

Table 13.3. Random-effects estimates and autoregressive component.

		$\hat{\sigma}_u^2$	$\hat{\sigma}_\delta^2$	$\hat{\sigma}_\xi^2$	$\hat{\sigma}_\eta^2$	$\hat{\gamma}$	$\hat{\rho}$
FM	waves 1-12	0.903	0.385	0.518	0.378	0.427	0.519
	waves 6-12	0.968	0.501	0.467	0.363	0.518	0.471
FS	waves 1-12 (linear trend)	0.846	0.494	0.351	0.242	0.585	0.557
	waves 6-12	0.943	0.505	0.437	0.392	0.536	0.322

All estimates are significantly different from zero at least at 1% level; for this reason we do not report p-values.

Let us now consider the random-effects for waves 6-12. In this case differences among parameter estimates are less evident except for the autoregressive parameter, which substantially seems to capture differences between the two measures.

13.5.2 The effect of covariates

The covariates considered in the model refer to household characteristics, and all of them are time dependent. The variables referring to the household head are: the age and the age square, AGE and AGE2; a dummy variable for the gender, SEX (1 if male); two dummies for the employment status, JOBSTATUS1 (1 if self or in paid employment) and JOBSTATUS2 (1 if unemployed); four dummies for educational level, EDUC1 (1 if first degree or more), EDUC2 (1 if HND, HNC2 or Teaching qualification), EDUC3 (1 if A level), EDUC4 (1 if O level); a dummy variable for the marital status, MARSTATUS (1 if married or in common law status); household size, HSIZE and HSIZE2 (size square); finally two dummies for macro regions, REGION3 and REGION4. REGION3 refers to Regions of the South West, West Midlands, Manchester and Merseyside; REGION4 refers to a large set of regions: region of the North West, Yorkshire, region of York & Humber, Tyne & Wear, region of the North, Wales and Scotland. The remaining areas are London (inner and outer, REGION1) and the South-East, East Anglia and East Midlands (REGION2).

All models have been estimated by maximum likelihood estimation using SAS PROC MIXED (Littell et al, 1996).

For each reference period (waves 1-12 and waves 1-6) two models have been estimated; one for the FM and one for the FS measure. Maximum likelihood estimates of the parameters are reported in Tables 13.4 and 13.6. In each model the dependent variable is the poverty indicator $y_{it}^{(\cdot)}$: thus a positive sign for a significant parameter corresponds to a higher deprivation risk, or more precisely to a higher membership function to the set of poor.

Let us first consider the longer reference period (waves 1-12). With regard to the time effect, considering the empirical evidence shown in Sect. 13.5.1, we specify a linear trend for the model based on the FS measure (variable TREND in Table 13.4), and use dummy variables for the time-effect in the model for the FM measure (variables DT2 – DT12 in Table 13.4). As expected, a decreasing level for both measures may be observed: this suggests a decreasing poverty risk from 1991 to 2002.

Let us now consider the effect of covariates. Observing Table 13.4 we note that for a subset of covariates included in the analysis there are more or less no differences between the FS and FM measures and the effects are as expected. In fact, the household head age has a quadratic effect on the degree of deprivation, with a minimum at about fifty years for the FM measure (this result is coherent with the lifecycle theory, see Modigliani, 1966), and at 55 years for the FS approach.

When compared to the reference category ("not to be in the labour force"), the poverty indicator is lower if the head of the household is employed or self-employed. According to the FM measure, the effect of the variable JOBSTATUS2 is, as expected, positive and therefore the poverty risk due to being unemployed is higher than the one for the reference category.

Table 13.4. Maximum likelihood estimates: waves 1-12

	FM measure		FS measure	
Fixed effects	Estimates	S.E.	Estimates	S.E.
Intercept	1.3284	0.043	2.9245	0.035
DT2	-0.1391	0.011		
DT3	-0.1656	0.013		
DT4	-0.2066	0.014		
DT5	-0.2516	0.014		
DT6	-0.3707	0.014		
DT7	-0.4276	0.014		
DT8	-0.4510	0.014		
DT9	-0.4715	0.015		
DT10	-0.5988	0.015		
DT11	-0.7805	0.015		
DT12	-0.9311	0.015		
TREND			-0.0724	0.001
AGE	-0.0303	0.002	-0.0769	0.001
AGE2	0.0003	0.000	0.0007	0.000
SEX	-0.0840	0.012	-0.0215	0.010
HHSIZE	0.0716	0.016	-0.4205	0.013
HHSIZE2	0.0065	0.023	0.0423	0.002
REGION3	0.1470	0.017	0.0839	0.015
REGION4	0.1463	0.016	0.0415	0.014
MARSTATUS	-0.2709	0.014	-0.1812	0.011
JOBSTATUS1	-0.3866	0.012	-0.1468	0.009
JOBSTATUS2	0.1028	0.018	-0.0287	0.013
EDUC1	-0.6498	0.020	-0.2497	0.017
EDUC2	-0.3106	0.023	-0.2791	0.019
EDUC3	-0.1891	0.017	-0.1742	0.014
EDUC4	-0.1205	0.015	-0.1403	0.012
-2log L	102515.1		76467.6	

By contrast, the effect of JOBSTATUS2 is not significantly different from zero in the case of the FS measure. The difference in the effects of JOBSTATUS2 in the two models is likely to be related to the volatility of the income in comparison with the possession of durable goods (we remind the reader that in this analysis the FS indicator is computed essentially by variables regarding housing conditions and possession of durable goods, see Table 13.2). The effect of the educational level of the household head is the same for the two measures: the degree of deprivation tends to decrease as the educational level increases. Married heads of household or in common law status make the membership function smaller than other marital statuses; such an effect is likely to be associated with the age of the head of household and/or with there being more than one wage earner in the household. According to both FM and FS measures a quadratic specification of the household size is significant. With regard to the FS measure, the membership function decreases with the increase of the household size up to five members (the minimum of the parabola). Where there are more than five members, it seems that there are not sufficient economic resources to meet the needs of the household members in terms of durable goods. On the contrary, monetary deprivation generally increases as the size of the household increases; it is reasonable to think that the increasing trend of the membership function is associated with the increasing number of children. The SEX variable is always significantly different from zero and its effect is negative; that is, households with a male household head are at an advantage. The degree of poverty is higher in Northern and Western regions than in the reference regions (Eastern regions and London).

Table 13.5. Components of variance; autocorrelated individual component models

		$\hat{\sigma}_u^2$	$\hat{\sigma}_\delta^2$	$\hat{\sigma}_\xi^2$	$\hat{\sigma}_\eta^2$	$\hat{\gamma}$	$\hat{\rho}$
FM	waves 1-12	0.687	0.232	0.455	0.355	0.337	0.470
	waves 6-12	0.714	0.298	0.416	0.344	0.417	0.416
FS	waves 1-12 (linear trend)	0.529	0.255	0.275	0.203	0.481	0.510
	waves 6-12	0.766	0.369	0.397	0.365	0.482	0.281

Let us now consider the parameters of the variance components for the analysis referring to waves 1-12, reported in Table 13.5. All the parameters are significantly different from zero. This result suggests that the effect of unobserved heterogeneity, interpreted as the effect of permanent differences among longitudinal units, plays an important role in the analysis of poverty dynamics. It is evident that parameter estimates in Table 13.5 are smaller than the ones in Table 13.3 due to the significant effect of missing covariates. Regarding the interpretation of these parameters we can refer to the remarks discussed in Sect. 13.5.1.

Finally, considering the parameter estimates using waves 6-12, reported in Table 13.6, the effect of the covariates does not change substantially with respect to the 1-12 waves analysis; the exceptions being the time effect, already observed in Sect. 13.5.1, and the regional effect. Obviously these differences concern only the model based on the FS measure, since the model is computed in the same way for both the analyses (waves 1-12 and waves 6-12). It is reasonable that in the Western Region the membership function to the set of poor decreases since the FS measure consists of a set of supplementary variables including environmental problems as well as life style factors.

Referring to the variance components (Table 13.5) the figures are similar to the ones already commented on in Sect. 13.5.1.

Table 13.6. Maximum likelihood estimates: waves 6-12

	FM measure		FS measure	
Fixed effects	Estimates	S.E.	Estimates	S.E.
Intercept	1.2035	0.056	2.5628	0.058
DT7	-0.0467	0.011	-0.1141	0.012
DT8	-0.0638	0.013	-0.1836	0.013
DT9	-0.0804	0.014	-0.1985	0.014
DT10	-0.1890	0.014	-0.2608	0.014
DT11	-0.3437	0.014	-0.3340	0.014
DT12	-0.4729	0.015	-0.3902	0.014
AGE	-0.0321	0.002	-0.0562	0.002
AGE2	0.0003	0.000	0.0005	0.000
SEX	-0.1016	0.015	-0.1714	0.016
HHSIZE	0.0668	0.021	-0.2618	0.022
HHSIZE2	0.0085	0.003	0.0361	0.003
REGION3	0.1513	0.021	-0.0686	0.022
REGION4	0.1650	0.019	-0.0165	0.020
MARSTATUS	-0.2614	0.018	-0.1870	0.019
JOBSTATUS1	-0.4132	0.015	-0.2036	0.016
JOBSTATUS2	0.1259	0.026	0.1018	0.028
EDUC1	-0.7075	0.025	-0.2431	0.026
EDUC2	-0.3527	0.029	-0.3096	0.030
EDUC3	-0.2140	0.021	-0.2154	0.022
EDUC4	-0.1537	0.019	-0.2069	0.020
-2log L	59475.7		62533.4	

13.6 Concluding remarks

In this Chapter we have proposed a methodology to model fuzzy and multidimensional poverty measures in order to study poverty dynamics; moreover we have analysed and discussed socio-demographic factors influencing those dynamics. We have taken into account two fuzzy measures, one based on a monetary variable only (FM) and the other based on supplementary variables (FS).

As far as the comparison between the two measures (FM and FS) is concerned, interesting results suggest that the FS measure can be used to complement the picture of poverty dynamics based on income, and that the simultaneous use of the two measures can help to understand the phenomenon of deprivation better.

We have also illustrated how fuzzy measures can overcome a further limitation of the traditional approach: overestimation of the mobility of the units near the poverty line.

From a methodological point of view, we are conscious that the specified model has the restrictions such as stationarity and interdependence between unobserved heterogeneity and covariates, which in principle could be relaxed. However, the relaxation of these restrictions would greatly increase the complexity of the estimation procedure. We also suspect that the improvements in the model specification do not necessarily improve the final results.

It is also important to point out that it would be interesting to consider dummy variables for household changes as covariate in the model, since these changes could influence the poverty process.

From an empirical point of view, we have observed that the number and mainly the nature of the supplementary indicators used for constructing the FS measure can greatly influence the model estimation results. This is particularly true for the model estimated for waves 6-12 which is based on a large number of heterogeneous items. For this reason, a further development of the analysis could consist in modelling several multidimensional poverty measures based on homogenous groups of indicators (dimensions) as proposed in Chapter 6, and in Sect. 7.3 of Chapter 7.

References

Anderson TW, Hsiao C (1981) Estimation of dynamic models with error components. Journal of the American Statistical Association 76:598–606

Anderson TW, Hsiao C (1982) Formulation and Estimation of Dynamic Models Using Panel Data. Journal of Econometrics 18:47–82

Bardasi E, Jenkins SP, Rigg J (2004) Documentation for derived current and annual net household income variables. BHPS 1–12, ISER unofficial supplement to BHPS data

Betti G, Cheli B, Cambini R (2004) A statistical model for the dynamics between two fuzzy states: theory and an application to poverty analysis. Metron 62:391-411

Betti G, D'Agostino A, Neri L (2002) Panel regression models for measuring multidimensional poverty dynamics. Statistical Methods and Applications 11:359-369

Betti G, Verma V (1999) Measuring the degree of poverty in a dynamic and comparative context: a multi-dimensional approach using fuzzy set theory. Proceedings of the ICCS-VI, Lahore, Pakistan, August 27–31, 1999, pp 289–301

Betti G, Verma V (2004) A methodology for the study of multi-dimensional and longitudinal aspects of poverty and deprivation. Università di Siena, Dipartimento di Metodi Quantitativi, Working Paper 49

Cerioli A, Zani S (1990) A fuzzy approach to the measurement of poverty. In: Dagum C, Zenga M (eds) Income and wealth distribution, inequality and poverty. Springer Verlag, Berlin, pp 272–284

Cheli B (1995) Totally Fuzzy and Relative Measures in Dynamics Context. Metron 53:83-205

Cheli B, Betti G (1999), Fuzzy Analysis of Poverty Dynamics on an Italian Pseudo Panel, 1985-1994. Metron 57:83-103

Cheli B, Lemmi A (1995) A "Totally" Fuzzy and Relative Approach to the Multidimensional Analysis of Poverty. Economic Notes 24:115-134

Devicenti F (2001) Poverty persistence in Britain: a multivariate analysis using The BHPS, 1991–1997. In: Moyes P, Seidl C, Shorrocks AF (eds) Inequalities: theory, measurement and applications. Journal of Economics Suppl. 9, pp 1–34

Goldstein H, Healy MJR, Rasbash J (1994) Multilevel time series models with applications to repeated measures data. Statistics in Medicine 13:1643–1655

Jenkins SP (2000) Modelling household income dynamics. Journal of Population Economics 13:529–567

Lillard LA, Willis RJ (1978) Dynamic aspect of earning mobility. Econometrica 46:985–1011

Littell RC, Milliken GA, Stroup WW, Wolfinger RD (1996) SAS System for Mixed Models. SAS Institute Inc., Cary, NC

Mansour H, Norheim EV, Rutledge JJ (1985) Maximum likelihood estimation of variance components in repeated measure designs assuming autoregressive errors. Biometrics 41:287–294

McClements LD (1977) Equivalence scales for children. Journal of Public Economics 8:191–210

Modigliani F (1966) The life cycle hypothesis of savings, the demand for wealth and the supply of capital. Social Research 33:160–217

Raghunathan TE, Lepkowski J, Van Voewyk J (2001) A multivariate technique for imputing missing values using a sequence of regression models. Survey Methodology 27:85–95

Stevens AH (1999) Climbing out of poverty, falling back in: measuring the persistence of poverty over multiple spells. Journal of Human Resources 34:557–588

Trivellato U (1998) Il monitoraggio della povertà e della sua dinamica: questioni di misura e evidenze empiriche. Statistica 58:549–575

Index